Crystallization
of
Membrane Proteins

Editor

Hartmut Michel

Director
Department of Molecular Membrane Biology
Max-Planck-Institute of Biophysics
Frankfurt/Main, Germany

CRC Press
Taylor & Francis Group
Boca Raton London New York

CRC Press is an imprint of the
Taylor & Francis Group, an **Informa** business

First published 1991 by CRC Press
Taylor & Francis Group
6000 Broken Sound Parkway NW, Suite 300
Boca Raton, FL 33487-2742

Reissued 2018 by CRC Press

Library of Congress Cataloging-in-Publication Data

Crystallization of membrane proteins / editor, Hartmut Michel.
     p.   cm.
   Includes bibliographical references and index.
   ISBN 0-8493-4816-1
   1. Membrane proteins. 2. Crystallization. I. Michel, Hartmut.
 QP552.M44C79   1990
 574.87'5-dc20   90-38256

A Library of Congress record exists under LC control number: 90038256

ISBN 13: 978-1-315-89217-7 (hbk)
ISBN 13: 978-1-351-07127-7 (ebk)

Visit the Taylor & Francis Web site at http://www.taylorandfrancis.com and the
CRC Press Web site at http://www.crcpress.com

PREFACE

The precise knowledge of the structure of biological macromolecules forms the basis of understanding their function and their mechanism of action. It also lays the foundation for rational protein and drug design. The only method to obtain this knowledge is still crystallography. At present, the structures of about 400 proteins are known at or nearly at atomic resolution. However, only two of them are membrane proteins or complexes of membrane proteins. The reason for the difference is not that crystals of membrane proteins present special problems when being analyzed. The reason is that membrane proteins resist forming well-ordered crystals. The intention of this book is to help to produce well-ordered crystals of membrane proteins and to provide guidelines. It is aimed at both biochemists and protein crystallographers.

The chapters in this book can be read independently. Chapter 1, written by Alexander McPherson, describes in detail the methods to crystallize soluble proteins. The same methods can be used to crystallize membrane proteins when the effects of the detergents (or surfactants) are taken into account. Therefore, Chapter 2, written by Martin Zulauf, deals intensively with the physicochemical aspects of detergents, as far as they are relevant for membrane protein crystallization. Chapter 3 describes the basic properties of membrane proteins and the general aspects of their crystallization, including some practical hints. Chapters 4, 5, and 6 contain more specific reports on individual examples.

All these contributors were able to describe at least one membrane protein crystal diffracting to high resolution, as listed in the final paragraph of Chapter 3. In Chapters 8 and 9 examples are presented, where the membrane proteins could be crystallized both in two and three dimensions due to special properties of the membrane proteins, but the two-dimensional crystals provided more information. Chapters 10 and 11 present two examples for the preparation of two-dimensional crystals. The methods used to obtain these two-dimensional crystals seem to be universally applicable. In general, the probability of obtaining two-dimensional crystals seems to be considerably higher than to obtain three-dimensional crystals. Finally, the appendices of this book contain lists of useful detergents including their relevant properties, and key references for the preparation and analysis of two-dimensional crystals of membrane proteins.

It is also my pleasure to express my thanks to Dieter Oesterhelt, without whose enthusiasm and support my personal success would not have been possible. I also sincerely thank Robert D. Gordon for help in editing this book and Bettina von Büdingen for engaged and valuable secretarial help.

Hartmut Michel
March 1990

THE EDITOR

Hartmut Michel is head of the department of molecular membrane biology and director at the Max-Planck-Institute of Biophysics at Frankfurt/Main, West Germany. At the same time he is an extraordinary professor of biochemistry at the University of Frankfurt.

Dr. Michel received a diploma in biochemistry at the University of Tübingen in 1975 and he obtained his Ph.D. in 1977 from the University of Würzurg. In Würzburg he started his experiments on the crystallization of membrane proteins in 1978, which became a success in 1981.

Dr. Michel is a member of the German Gesellschaft für Biologische Chemie, the Gesellschaft Deutscher Chemiker, the Gesellschaft für Physikalische Biologie, the Max-Planck-Society (scientific member), the European Molecular Biology Organization, the International Academy of Science, and the European Academy of Arts, Sciences and Humanities.

For the crystallization of membrane proteins and the determination of the three-dimensional structure of the photosynthetic reaction center from the purple bacterium *Rhodopseudomonas viridis*, he received among other prizes the Biophysics Prize of the American Physical Society together with Johann Deisenhofer, the Otto-Klung-Preis for the "best German chemist under age 40", the Leibniz-Preis of the Deutsche Forschungsgemeinschaft, and the Nobel Prize for Chemistry in 1988 together with Johann Deisenhofer and Robert Huber.

Dr. Michel's current major research interest is still to understand the function of membrane proteins on the basis of their structure. Towards this goal the crystallization of membrane proteins still is the key.

CONTRIBUTORS

James Paul Allen, Ph.D.
Assistant Professor
Department of Chemistry
Arizona State University
Tempe, Arizona

Richard J. Cogdell, Ph.D.
Professor
Department of Botany
University of Glasgow
Glasgow, Scotland

Deborah J. Dawkins, Ph.D.
Research Fellow
Department of Botany
University of Glasgow
Glasgow, Scotland

George Feher, Ph.D.
Professor
Department of Physics
University of California
La Jolla, California

Linda A. Ferguson, B.Sc.
Department of Botany
University of Glasgow
Glasgow, Scotland

R. Michael Garavito, Ph.D.
Assistant Professor
Department of Biochemistry
 and Molecular Biology
University of Chicago
Chicago, Illinois

Werner Kühlbrandt, Ph.D.
Group Leader
Department of Biological Structures
European Molecular Biology Laboratory
Heidelberg, Germany

Kevin R. Leonard, Ph.D.
Group Leader
Department of Biological Structures
European Molecular Biology Laboratory
Heidelberg, Germany

Anthony N. Martonosi, M.D.
Professor
Department of Biochemistry
SUNY Health Science Center
Syracuse, New York

Alexander McPherson, Ph.D.
Professor and Chairman
Department of Biochemistry
University of California
Riverside, California

Hartmut Michel, Dr. rer. nat.
Director and Professor
Department of Molecular Membrane
 Biology
Max-Planck-Institute of Biophysics
Frankfurt/Main, Germany

Kenneth R. Miller, Ph.D.
Professor of Biology
Division of Biology and Medicine
Brown University
Providence, Rhode Island

Slawomir Pikula, Ph.D.
Research Scientist
Nencki Institute
Polish Academy of Sciences
Warsaw, Poland

Kenneth A. Taylor, Ph.D.
Associate Professor
Department of Cell Biology
Duke University Medical Center
Durham, North Carolina

Thomas Wacker, Dr. rer. nat.
Institute of Biophysics and Radiology
University of Freiburg
Freiburg, Germany

Hanns Weiss, Dr. med.
Professor
Institute of Biochemistry
University of Dusseldorf
Düsseldorf, Germany

Wolfram Welte, Dr. rer. nat.
Institute of Biophysics and Radiology
University of Freiburg
Freiburg, Germany

Kevin J. Woolley, Ph.D.
Research Fellow
Department of Botany
University of Glasgow
Glasgow, Scotland

Martin Zulauf, Dr. phil. nat.
Central Research Units
Hoffmann-LaRoche & Co.
Basel, Switzerland

TABLE OF CONTENTS

Chapter 1

USEFUL PRINCIPLES FOR THE CRYSTALLIZATION OF PROTEINS

Alexander McPherson

TABLE OF CONTENTS

I. SOLUBILITY AND SUPERSATURATION

A solution is formed as a consequence of dissolving a solute in a solvent in specific proportions. If the quantity of solvent is fixed, then at a given temperature the toal amount of solute that can enter the liquid phase is strictly limited. The limit is determined by the physical and chemical properties of the solvent and solute and the way they interact with one another. The maximum amount of solute that can be dissolved under constant conditions of temperature, pH, pressure, etc., is defined as the solubility of the compound. Once the solubility limit has been reached then additional added solute does not lead to any increase in the concentration of solute molecules in the solution.

Two phases exist in equilibrium, the solid phase and the liquid phase. Molecules may leave the solid phase and enter solution, but only at a rate equal to that at which other molecules leave the solution to join the solid phase. In this state of equilibrium, where the solubility limit of the solute has been reached and the two phases are balanced, the solution is said to be saturated.

At saturation, or at equilibrium, no increase in the proportion of solid phase can occur since it would be matched by a countervailing dissolution. Hence crystals cannot grow from a saturated solution. They do not form and grow from an equilibrium state. For a crystal to appear and grow, the solution must be in a nonequilibrium state. There must be a driving force toward the establishment of equilibrium. The appearance of crystals, like those shown in Figures 1 and 2, and their growth is one manifestation. Clearly then, if one wishes to grow crystals of any sort, an objective must be to remove the system to some nonequilibrium state and utilize its return to force the exclusion of molecules into the solid state, the crystal.

Assume that a solute has been dissolved to the limit of its solubility and the saturated state prevails. If solvent is then withdrawn by slow evaporation, if the temperature is reduced, or some other property of the system is changed then more solute will be in solution in proportion to solvent then the solubility should allow. This condition is known as supersaturation. If solid is introduced as a second phase, then equilibrium will be re-established as molecules leave the solvent to join the solid phase until strict saturation is reached.

If no second phase is present, as conditions are changed, then solute will not immediately partition into two phases, and the solution will remain in the non-equilibrium state, the supersaturated state. No second phase occurs spontaneously as the saturation limit is exceeded because energy, analogous to the activation energy of a chemical reaction, is required to create the second phase, the stable nucleus of a crystal or a precipitate. Thus a kinetic or energy barrier allows conditions to proceed further and further from equilibrium, into the zone of supersaturation. The line indicative of saturation is thus also a demarcation that marks the requirement for energy requiring events to occur that will lead to a second phase, the nucleus of a crystal or the nonspecific aggregate that characterizes a precipitate. For a clear and detailed discussion of supersaturation as it pertains to crystal growth, the classic work of Buckley should be consulted.[12]

Once a stable nucleus forms on the supersaturated side of the saturation limit it will continue to grow so long as the system has not regained equilibrium, that is, returned to the saturated state. While non-equilibrium forces prevail and some degree of supersaturation exists, a crystal will grow or precipitate will continue to form.

A crucial term is "stable nucleus", for this in fact represents the second state, the solid phase, the crystal. Many aggregates or nuclei spontaneously form once supersaturation is achieved, but most are not "stable". Instead of continuing to develop, they redissolve as fast as they form and their constituent molecules return to solution. A "stable nucleus" is a molecular aggregate of such size and physical coherence that it will enlist new molecules into its growing surfaces faster than others that are lost to solution, i.e., it will continue to grow so long as the system is supersaturated.

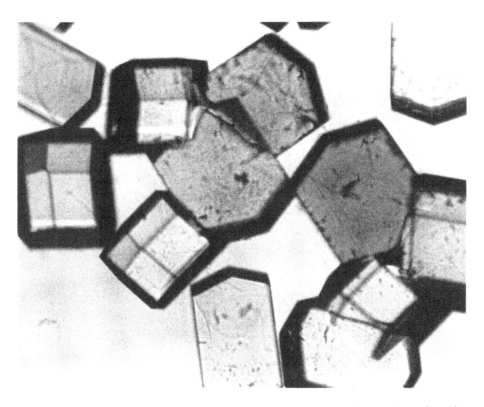

FIGURE 1.   Tetragonal crystals of lysozyme, one of the more readily crystallized proteins and a subject of much crystallization research are seen here under a low power light microscope.

In classical theories of growth of conventional crystals, the region of supersaturation that pertains above saturation is further divided into the metastable region and the labile region as shown in Figure 3.[69,70,78] It is commonly agreed that stable nuclei cannot form in the metastable region just beyond saturation. If a stable nucleus is present in the metastable region, then it can and will continue to grow. The labile region of supersaturation is discriminated from the metastable in that stable nuclei can spontaneously form. They will accumulate molecules and thus deplete the liquid phase until it returns to the metastable state and ultimately to the saturated equilibrium state.

An important point, shown graphically in Figure 3, is that there are two regions above saturation, one of which can support crystal growth but not formation of stable nuclei, and the other which can yield nuclei as well as support growth. Now the rate of crystal growth is some function of the distance of the solution from the equilibrium position at saturation, thus a nucleus that forms far from equilibrium and well into the labile region will grow very rapidly at first and, as the solution is depleted and moves back toward the metastable state, it will grow slower and slower. The nearer the system is to the metastable state when a stable nucleus forms, then the slower it will proceed to mature.

It might appear that the best approach for obtaining crystals is to press the system as far into the labile region, supersaturation, as possible. There, the probability of nuclei formation is greatest, the speed of growth is greatest, and the likelihood of crystals is maximized. As the labile region is penetrated further, however, the probability of spontaneous and uncontrolled nucleation is also enhanced. Thus crystallization from solutions in the labile region far from the metastable state frequently results in extensive and uncontrolleld "showers" of crystals. By virtue of their number, none is favored and, in general, none will grow to a size suitable for X-ray diffraction studies. In addition, when crystallization is initiated

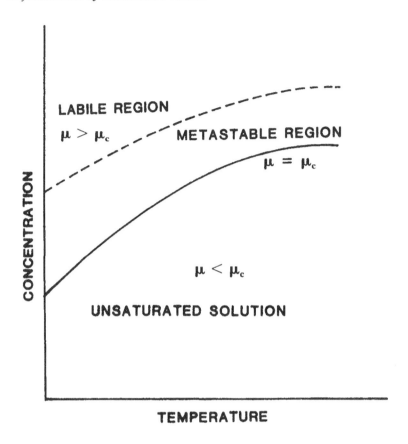

FIGURE 3.   A typical solubility diagram showing the labile, metastable, and un-
saturated regions. The solid line represents the saturation boundary. Although this
diagram illustrates the case for concentration as a function of temperature, other
parameters which effect solubility may be similarly treated. In this diagram, μ_c is the
chemical potential of a crystal; μ_1 is the chemical potential of the same crystal in
solution. The diagram is after Petrov.[36]

enhance the likelihood of nuclei formation and growth, then one must do whatever is possible
to insure the greatest number of the most stable bonds between the solute molecules in the
solid state.

We may ask why molecules should arrange themselves into perfectly ordered and periodic
crystal lattices, exemplified by those in Figure 4, when they could equally well form disordered aggregates, which we commonly call precipitates. The answer is again the same as
for the question as to why they leave the solution phase at all, to form the greatest number
of most stable bonds; to minimize the potential energy of the system.[97] It can be shown
mathematically, and it is intuitively so as well, that the optimal bonding arrangement that
a system can achieve is reached when a single, optimal bonding pattern between individual
molecules is utilized repeatedly in a three dimensionally periodic manner. While precipitates
represent, in general, a low energy state for solute in equilibrium with a solution phase,
they are not the state of lowest free energy. Precipitates are false minima on an energy
surface.

A frequently observed phenomena is the formation of a precipitate followed by its partial
or total dissolution concomitant with the formation and growth of crystals. The converse is
never observed. This is an empirical proof that the crystal represents a more favorable energy
state. Occasionally, and particularly with proteins, crystals of one morphology or internal
structure may grow and, as these are seen to slowly redissolve, a completely different crystal

FIGURE 2. Large rhombohedral crystals of canavalin from Jack Bean, a 150,000 D trimeric molecule comprised of three identical 50,000 D subunits. The ease with which this protein can be crystallized has made it a favored subject for crystallization research including experiments in microgravity aboard the U.S. Space Shuttle. The crystals shown here approach 1 mm in their longest dimension.

from a point of high supersaturation, then initial growth is extremely rapid. Rapid growth is frequently associated with the occurrence of flaws and defects. Hence crystals produced from extremely saturated solutions tend to be numerous, small, and afflicted with growth defects.

In terms of the phase diagram, ideal crystal growth would begin with nuclei formed in the labile region but just beyond the metastable. There, growth would occur slowly, the solution by depletion would return to the metastable state where no more stable nuclei could form, and the few nuclei that had established themselves would continue to grow to maturity at a pace free of defect formation. Thus in growing crystals for X-ray diffraction analysis, one attempts, by either dehydration or alteration of physical conditions, to transport the solution into a labile, supersaturated state, but one as close as possible to the metastable phase.

II. WHY DO CRYSTALS GROW?

The natural inclination of any system proceeding toward equilibrium is to maximize the extent of disorder, by freeing individual constituents from physical or chemical constraint. At the same time, there is a necessity to minimize the potential energy of the system. This occurs by the formation of chemical bonds and interactions which generally provide negative free energy. Clearly the asembly of molecules into a fixed lattice severely reduces their mobility and freedom, yet crystals do form and grow.

It follows that crystal formation and growth must therefore be a consequence of chemical and physical bonds arising in the crystalline state that either cannot be formed in solution or are stronger than those that can. These bonds are, in fact, what hold crystals together. They are the energetically favorable intermolecular interactions that drive crystal growth in spite of the resistance to molecular constraint. From this it is clear that if one wishes to

FIGURE 4.   Thick orthorhombic plates of canavalin are seen intermeshing with one another. The crystals are more than 2 mm in their longest dimension and each originates from a unique nucleus.

form appears and grows. Here the two different crystal forms, or polymorphs, represent two distinct but very similar energy states. The latter has but a marginal advantage in energy terms on the former. This phenomenon is illustrated in Figures 5 and 6.

III. PROTEINS PRESENT SPECIAL PROBLEMS FOR CRYSTALLOGRAPHERS

In principle, the crystallization of a protein, nucleic acid, or virus is a little different than the crystallization of conventional small molecules (for a discussion of the similarities and differences see Reference 32 and 84). It requires the gradual creation of a supersaturated solution of the macromolecule followed by spontaneous formation of crystal growth centers or nuclei. Once growth has begun, emphasis changes to maintenance of essentially invariant conditions so as to sustain the continued ordered addition of single molecules, or perhaps ordered aggregates, to the surfaces of the developing crystal. Under nonequilibrium conditions of supersaturation, the system is driven toward a final state in which the solute is partitioned between a soluble and solid phase. Although the individual macromolecules lose rotational and translational freedom, thereby lowering the entropy of the system, they, at the same time, form many new, stable chemical bonds. This reduces the potential or free energy of the system and provides the motivation for the self-ordering process.

The greater difficulties that arise in the crystallization of macromolecules in comparison with conventional small molecules stem from the greater complexity, lability, and dynamic properties of proteins and nucleic acids (for excellent texts on the properties of such macromolecules see Reference 18 and 86). The rather simple explanation offered above of labile and metastable regions of supersaturation are still applicable to macromolecules, but it must

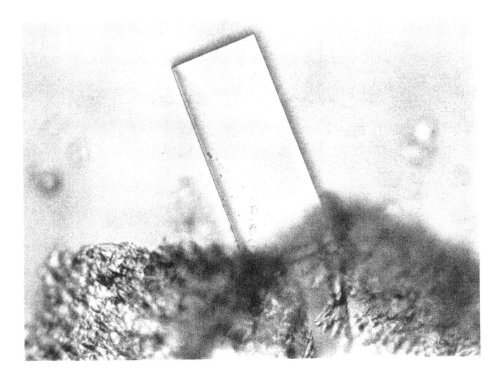

FIGURE 5. The gray mass at the bottom in this photomicrograph is a large rhombohedral crystal of canavalin that is slowly degrading and dissolving. The orthorhombic crystal seen at the center is growing directly from this mass of crystalline protein and will eventually consume it.

now be borne in mind that as conditions are adjusted to transport the solution away from equilibrium by alteration of its physical and chemical properties, the very nature of the solute molecules is changing as well. As temperature, pH, pressure, or solvation are changed, so are the conformation, charge state, or size of the solute macromolecules.

In addition, proteins and nucleic acids are very sensitive to their environment, and if exposed to sufficiently severe conditions, may denature, degrade, or randomize in a manner that ultimately precludes any hope of their forming crystals. They must be constantly maintained in an essentially hydrated state at or near physiological pH and temperature. Thus common methods for the crystallization of conventional molecules such as evaporation of solvent, dramatic temperature variation, or addition of strong organic solvents are unsuitable and destructive. They must be supplanted with more gentle and restricted techniques.

Certain properties of proteins render the problem of crystallization particularly difficult to treat systematically or in strictly analytical terms. For example, all proteins are different from one another. Each has a unique chemical composition, polypeptide chain folding and three-dimensional structure. The surface of a protein is a mixture of positive and negative ionizable groups that alter their electrostatic properties as a function of pH or electrolyte. Proteins, furthermore, are not static molecules but vibrate and writhe in solution and generally assume more than one discrete conformational state. Proteins are very large molecules ranging from a few thousand daltons molecular weight to hundreds of thousands of daltons. Their shapes are irregular, and though generally compact, exhibit occlusions and protrusions that make close packing awkward. Some of the biologically most important are not just composed of amino acids but are conjugated to lipids or carbohydrate chains as well. Finally, proteins have shells of bound solvent molecules held to them with various affinities that shield and modify many of the proteins inherent characteristics. Yet proteins, even with complicated features such as those in Figure 7 and 8, do crystallize.

FIGURE 6. An example of crystal polymorphism is provided here by the protein canavalin from Jack Beans. In the same sample both the hexagonal prism habit (space group P6$_3$) and the orthorhombic form (space group C222$_1$) appear. In general the hexagonal prisms grow first but ultimately dissolve to the benefit of the more stable orthorhombic crystals.

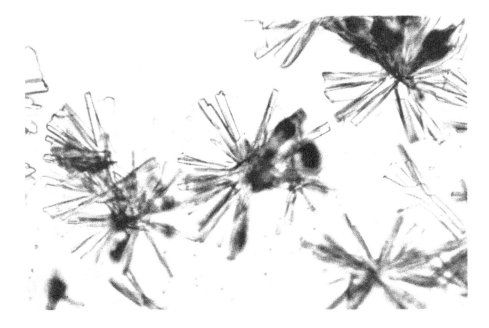

FIGURE 7. Crystals of the α subunit of the leutinizing hormone protein grown from PEG 4000. The protein is a glycoprotein isolated from bovine pituitary tissue and is seen to be quite heterogeneous on isoelectric focusing gels.

A

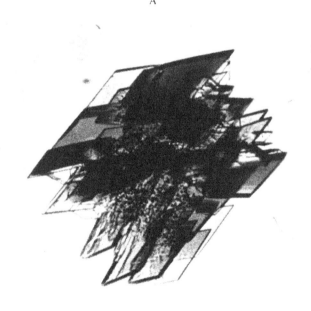

B

FIGURE 8. Two examples of glycoproteins that contain large proportions of covalently bound carbohydrate but which crystallize are in (A) α_1-acid glycoprotein from human serum (63) and (B) the cortisol transport protein from guinea pig serum (68a). α_1-Acid glycoprotein and CBG contain 49% and 29% carbohydrate by weight and are crystallized from 65% $(NH_4)_2SO_4$ and 14% PEG 4000, respectively.

IV. THE UNIQUE FEATURES OF MACROMOLECULAR CRYSTALS

Biomolecular crystals are composed of approximately 50% solvent on the average, though this may vary from 30 to 90% depending on the particular macromolecule.[57] The protein or nucleic acid occupies the remaining volume so that the entire crystal is in many ways an ordered gel with extensive interstitial spaces through which solvent and other small molecules may freely diffuse. This high solvent content is shown strikingly by the electron micrographs of protein crystals seen in Figure 9.

In proportion to molecular weight, the number of interactions (salt, hydrogen, hydrophobic) that a conventional molecule forms in a crystal with its neighbors far exceeds the very few exhibited by crystalline macromolecules. Since these contacts provide the lattice forces that maintian the integrity of the crystal, this largely explains the difference in properties between crystals of small molecules and macromolecules as well as why it is so difficult to grow protein crystals.

Because proteins are sensitive and labile macromolecules that readily lose their native structures, the only conditions that can support crystal growth are those that cause no perturbation of the molecular properties. Thus protein crystals, maintained within a narrow range of pH, temperature and ionic strength, are grown from a solution to which they are tolerant, called the mother liquor. As complete hydration is essential for the maintenance of structure, protein crystals are always, even during data collection, bathed in the mother liquor.[83]

Although morphologically indistinguishable, there are important differences between crystals of low molecular weight compounds and crystals of proteins and nucleic acids. Crystals of small molecules exhibit firm lattice forces, are highly ordered, are usually mechanically hard and easy to manipulate, can be exposed to air, have strong optical properties, and diffract X-rays intensely. Macromolecular crystals are by comparison usually much more limited in size, are very soft and crush easily, disintegrate if allowed to dehydrate, exhibit weak optical properties, and diffract X-rays poorly. Macromolecular crystals are temperature sensitive and undergo extensive damage after prolonged exposure to radiation. In general, many crystals must be analyzed for a structure determination to be successful.

The extent of the diffraction pattern from a crystal is directly correlated with its degree of internal order. The more extensive the pattern, or the higher the resolution to which it extends, then the more uniform are the molecules in the crystal and the more precise is their periodic arrangement. The level of detail to which atomic positions can be determined by a crystal structure analysis corresponds closely with the degree of crystalline order. While conventional molecular crystals often diffract almost to their theoretical limit of resolution, protein crystals by comparison are characterized by diffraction patterns of limited extent.

The liquid channels and solvent cavities that characterize macromolecular crystals are primarily responsible for the limited resolution of the diffraction patterns. Due to the relatively large spaces between adjacent molecules, seen clearly for example in Figure 10, and the consequent weak lattice forces, every molecule in the crystal may not occupy exactly equivalent translational positions in the crystal but they may vary slightly in mean position from lattice point to lattice point. Furthermore, protein molecules in a particular crystal may exhibit slight variations in the course of their polypeptide chains or the dispositions of side groups, because of their complexity and their potential for conformational dynamics.

Although the presence of extensive solvent regions is a major contributor to the poor quality of protein crystals, it is also responsible for their value to biochemists. The individual macromolecules in protein crystals are surrounded by hydration layers that maintain their structure virtually unchanged from that found in bulk solvent. As a result, ligand binding, enzymatic and spectroscopic characteristics, and other biochemical features are essentially

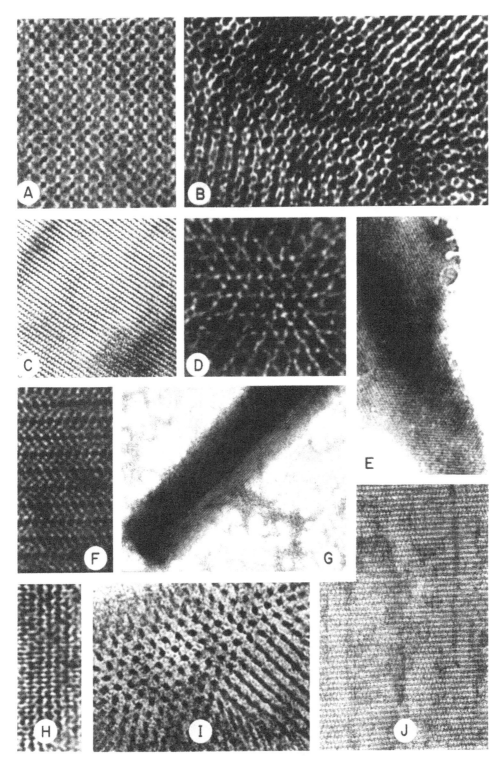

FIGURE 9. An array of electron micrographs of a variety of negatively stained protein microcrystals. In (A) is orthorhombic canavalin from Jack Bean, (B) orthorhombic crystals of pig pancreas α-amylase, and (C) a second crystal form of the same protein. In (D) is seen hexagonal canavalin crystals, in (E) the lectin from abrus precatorius, (F) is rhombohedral canavalin, and (G) is a microcrystal of orthorhombic B. Subtilis α-amylase. In (H) is hexagonal concanavalin B from Jack Bean, (I) is trigonal beef liver catalase, and (J) is the orthorhombic crystal form of the same protein. (From McPherson, A., *The Preparation and Analysis of Protein Crystals*, John Wiley & Sons, New York, 1982. With permission.)

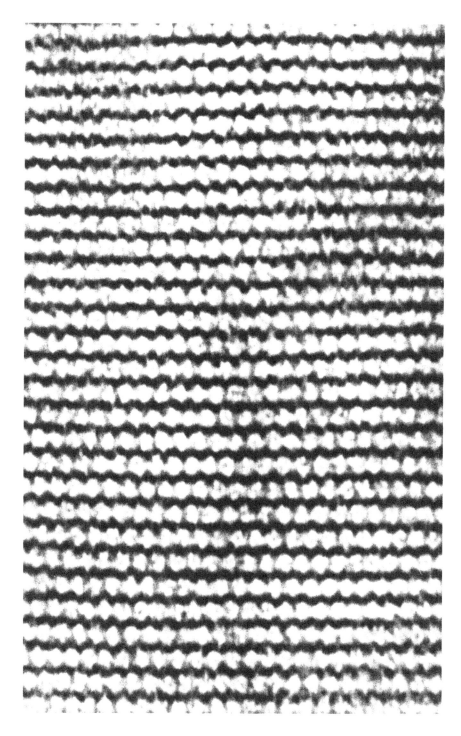

FIGURE 10.   An electron micrograph of a microcrystal of pig pancreas α-amylase negatively stained with uranyl acetate. The light colored, oval units are composed of two molecules of α-amylase each having a molecular weight of 50,000 D. Note the order that remains in the crystal and the clarity of the protein molecules even after total dehydration and heavy metal staining. Note also the high proportion, nearly 50%, of the crystal that would be occupied by solvent, here replaced by the uranyl acetate.

the same as for the native molecule in solution. In addition, the size of the solvent channels is such that small compounds, which may be ions, ligands, substrates, coenzymes, inhibitors, drugs, or other effector molecules, may be freely diffused into and out of the crystals. Crystalline enzymes, though immobilized, are completely accessible for experimentation through alteration of the surrounding mother liquor. Thus, a protein crystal can serve as a veritable ligand binding laboratory. For numerous examples of the use of protein crystals for the elucidation of enzyme mechanisms see Jurnak and McPherson.[40]

Protein crystals, because of their unique properties can permit experiments that are impossible with crystals of small, conventional molecules. Enzymes, which are proteins, can in the crystalline state bind specific substrates or coenzymes and demonstrate significant catalytic efficiency in converting them to product. Though the rate of catalysis is markedly decreased as a result of diffusion limitations, the phenomenon demonstrates clearly that enzyme molecules immobilized in a crystal lattice still possess the necessary mobility and conformational freedom to carry out their intended task.[81] It also points out that the bonds between protein molecules in the lattice must indeed be few and weak to permit such relatively unconstrained movement.

Another interesting feature of protein crystals is that their constituent molecules can be crosslinked *in situ*.[27,81] This can be accomplished, for example, by exposing them to a 3% glutaraldehyde solution for a few hours. Glutaraldehyde, which has two free aldehyde groups, will react with lysine residues on adjacent protein molecules in the lattice, effectively chaining them to one another. Such crosslinked crystals are much harder and far less fragile than non-crosslinked equivalents and far easier to deal with in the mechanical sense. Unfortunately, the protein molecules become statistically disordered in the process and the diffraction quality of fully crosslinked crystals is usually negligible.

Protein crystals may also be exposed to glutaraldehyde for a brief period only. This results in the crosslinking of the molecules in the layers at and near the surface of the crystals. These "surface crosslinked" crystals can then be used for a variety of interesting studies. Haas (personal communication), for example, showed that when a surface crosslinked crystal was transferred from its normal mother liquor to pure water, the surface features of the crystal remained fully intact while the interior completely dissolved away. All that remained were transparent casings of several protein molecules thick having the exact habit of the initial crystal. This experiment strikingly demonstrated that protein crystals must have pores in their surfaces and internal channels sufficiently large for the constituent protein molecules themselves to diffuse in and out.

Protein crystals can, under some conditions, be frozen and maintained at very low temperatures while they are examined by X-ray diffraction analysis.[56] This cannot, in general, be done simply by freezing a native-protein crystal because the formation of ice in the interior of the crystals destroys its order and renders it useless. Haas, however, showed that protein crystals could be perfused with 3 M sucrose and rapidly frozen in liquid nitrogen without damage.[28] The sucrose formed a non-crystalline, innocuous glass which inflicted no damage on the protein crystals. Haas was further able to show that these sucrose perfused protein crystals when maintained at cryogenic temperatures could sustain several hundred hours of X-ray exposure without measureable decay, far more than the equivalent native crystals. Others have extended these cryogenic studies on protein crystals and shown that X-ray studies could be performed on stable enzyme substrate intermediates that were only transient in native crystals at room temperature.

Protein crystals may also be examined and analyzed using electron microscopy as well as X-ray diffraction analysis. Microcrystals of many proteins can be spread on grids, negatively stained with uranyl or lead and subjected to otherwise conventional procedures. Examples of the results are shown in Figures 9 and 10. Becaues of the damaging effects of the stain and the trauma of dehydration in the vacuum chamber of the microscope, the lattice is usually severely disordered. Nonetheless, useful images of crystalline molecular arrays can fequently be obtained that display a resolution of detail on the order of 10 Å.

More recently the techniques of low dose transmission electron microscopy combined with electron diffraction have been utilized to study the structures of protein microcrystals.[37,50,96] Usually performed in the presence of a cryogen and at low temperatures, these investigations can in some instances provide low resolution (3 to 5 Å) images of the protein molecules that make up the crystals. Advancements in the application of these techniques could do much to relieve the necessity for the production of large, perfect crystals for X-ray diffraction analysis.

V. CRYSTALLIZATION: THE STRATEGY

The most common approach to crystallizing macromolecules is to gradually alter the characteristics of a highly concentrated protein solution. The changes are designed to deprive the protein molecules of sufficient ions or water to maintain hydration, to significantly disrupt their hydration layers, or to exclude them in some way from the bulk solvent. An alternative approach is to decrease the dielectric properties of the solution by the addition of organic solvents so as to reduce the effective electrostatic shielding between macromolecules. In both cases no effort must be spared in adjusting the parameters of the system, solvent and solute, to encourage and promote specific bonding interactions between molecules or to stabilize them once they have formed. This latter aspect of the problem generally depends on the molecular properties of the particular protein or nucleic acid being crystallized.

The strategy employed to bring about crystallization is to guide the system very slowly toward a state of reduced solubility by modifying the properties of the solvent through equilibration with precipitating agents or by altering some physical property, such as pH, thus achieving a limited degree of supersaturation. In extremely concentrated solutions the molecules may aggregate as an amorphous precipitate. This state is to be avoided if possible and is one indicative that saturation has proceeded too extensively or too rapidly. One must endeavor to approach very slowly the point of inadequate solvation and thereby allow the macromolecules sufficient opportunity to order themselves in a crystalline lattice. At the same time, the component variables of the system must be initially set or gradually modified to ensure that the macromolecules will take advantage of the greatest number of favorable interactions with neighbors. For a thorough discussion of protein solubility in a number of solutions see Tsutomu and Timasheff.[95]

VI. SALTING-OUT AND SALTING-IN

The most common procedure for inducing proteins to separate from solution and enter the solid state is to gradually increase the level of saturation of a salt. Traditionally the salt has been ammonium sulfate, but others (see section on precipitants) are also in wide use. Usually, the protein separates as a precipitate, but with appropriate care, increase in salt concentration can be used to grow protein crystals. At this writing, in fact, this approach has probably yielded more varieties of protein crystals than any other.

For a specific protein, the precipitation points or solubility minima usually depend critically on the pH, temperature, the chemical composition of the precipitant, and the properties of both the protein and the solvent. In Figure 11 the plot of log solubility at constant temperature and pressure is shown for a typical protein as a function of salt concentration. At very low ionic strengths a phenomenon known as "salting-in" occurs in which the solubility of the protein increases as the ionic strength increases. The physical effect that induces reduced solubility at very low ionic strength, is the removal of ions essential for satisfying the electrostatic requirements of the protein molecules. As these ions,

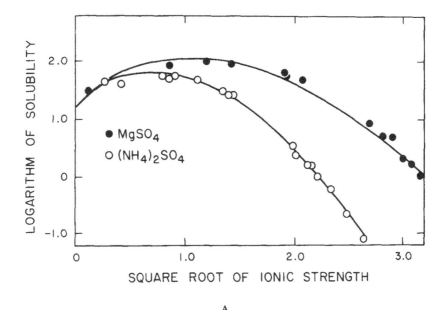

A

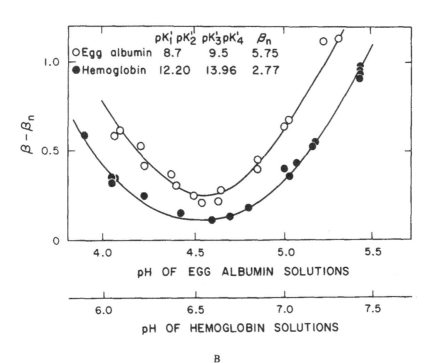

B

FIGURE 11.   The solubility behavior of typical proteins as a function of ionic strength and pH. In (A) the solubility is a function of ionic strength produced by two different salts $MgSO_4$ and $(NH_4)_2SO_4$ for the protein enolase. In (B) the solubility of two proteins, ovalbumin and hemoglobin is a function of pH at constant ionic strength. The two end points of the curves in (A) correspond to the "salting in" and "salting out" regions of the solubility diagram. In (B) the minima correspond to the isoelectric points of the proteins.

and in this region of solubility cations are most important,[19] are removed the protein molecules seek to satisfy their electrostatic requirements through interactions among themselves. Thus, they tend to aggregate and separate from solution. An alternative description is to say that the activity coefficient of the protein is reduced at very low ionic strength.

The salting-in effect, when applied in the direction of reduced ionic strength can itself be used as a crystallization tool. In practice, one simply dialyzes a protein that is soluble at high ionic strength extensively against distilled water. Many proteins such as catalase, concanavalin B, and a host of immunoglobulins and seed proteins have been crystallized by this means.[60,61]

As ionic strength is increased in Figure 11 the solution again reaches a point where the solute begins to separate from solvent and preferentially form self interactions that result in crystals or precipitate. The explanation for this phenomenon is that salt ions compete for the chemical attention of the solvent molecules, that is, water. Both the salt ions and the protein molecules require hydration layers to maintain their solubility. When competition between ions and protein becomes sufficiently intense, the protein molecules begin to self associate in order, by intermolecular interactions, to satisfy their electrostatic requirements. Thus, dehydration, or the elimination or perturbation of solvent shells around protein molecules causes them to become insoluble.

An analytical expression suggested initially by Hofmeister for conventional small molecules[33] and later refined by Cohn and his colleagues[16,17] has been used to describe the effects of pH, temperature, salt concentration and the nature of the protein on the equilibrium between the soluble and insoluble phases. This was promoted by the observation that if the logarithm of the solubility of a specific protein in the "salting-out" region at fixed temperature and pH is plotted against the salt concentration, a nearly linear relationship is obtained.

In this equation

$$\log S = \beta - K_s (I/2)$$

S is the solubility of the protein in grams per liter of solution, $I/2$ is the ionic strength in moles per liter of solution and β and K_s are constants. The salting out constant K_s is independent of temperature and pH and is a function of the nature of the salt ions and of the protein. The efficiency of a particular salt is proportional to the square of the number of charges it carries that can be invovled in the binding of water molecules since ionic strength $I/2 = mz^2$ where m is the concentration of each ion in solution and z its valence number. Thus, divalent and trivalent ions, such as sulfate and phosphate, are most commonly employed. The constant β represents the logarithm of the solubility extrapolated to an ionic strength of zero and it is dependent on temperature and hydrogen ion concentration.

The theory on which the Cohn and Ferry equation is based does not take into account changes in the character of the protein as pH or temperature is varied. In practice, a specific protein at constant salt concentration may exhibit several solubility minima as the pH is altered and this is also seen in Figure 11. It is nevertheless useful to divide the separation of protein from solution as crystals or precipitate according to methods based on either variation of ionic strength at constant pH and temperature, K_s separation, or based on variation of pH or temperature at constant precipitant concentration, β separation. A more precise and detailed discussion of the mathematical treatment of protein solubility may be found in references 19 and 95.

It should be noted in passing, that the principles described here for salting out with a true salt, that is K_s and β separation, are not appreciably different if precipitating agents such as PEG are used instead. In practice, proteins may equally well be crystallized from solution by increasing the PEG concentration at constant pH and temperature, or at constant

PEG concentration by variation of pH or temperature. This holds true for the most part with other organic precipitants such as MPD as well.

VII. THE STAGES OF CRYSTALLIZATION

There are, in the process of obtaining large protein crystals, three distinct but more or less continuous stages; (1) formation of stable nuclei, (2) growth of nuclei to form mature crystals, and (3) termination of growth.[42] An important point to recognize, is that the conditions for most probable nucleation and optimal growth are not in general the same. We tend to assume that once visible microcrystals appear then the best practical course to follow is to leave them completely undisturbed. In fact, it may well be that appearance of microcrystals should be the signal for change to other conditions more suitable for growth. This is equivalent to saying that once the labile region of supersaturation has been penetrated sufficiently to yield the essential stable nuclei, the best course is then to return into the metastable region as close to saturation as possible.

Manipulation of ambient conditions to optimize growth conditions for individual crystallization samples of a vast array of trials is currently outside the range of our abilities and understanding. It is encouraging, however, that investigations are now underway and experimental apparatus being designed in several laboratories to accomplish precisely this objective.

Prior to nucleation, monomers of protein are in equilibrium with various kinds of aggregates. In the metastable region of supersaturation the aggregates are unstable and dissolve as rapidly as they form, while in the labile region they continue to develop in size. The nuclei, as Kam and Feher have pointed out, may be of two fundamental kinds.[42] They may be linear or branched oligomers displaying growth in one or two dimensions, or they may be ordered aggregates with monomers adding in all three directions. While the former ultimately give rise to precipitate, the latter eventually represent the solid state as true crystals.

It is unfortunately true that it requires a greater activation energy for protein monomers to join a growing, ordered, three dimensional aggregate, a crystal nucleus, than to enlist in the formation of a linear chain. The counter point of course is that once in place, the monomer integrated into a true crystal is the more stable of the two. The height of the activation barrier for addition of monomers to an aggregate will be a complex function of the many parameters of the system that describe quantitatively the protein-protein and protein-solvent interactions.

Even before visible crystal nuclei or precipitate are formed, the distribution of aggregate types and sizes may be monitored using physical techniques such as inelastic light scattering.[42] The variety of ordered three dimensional aggregates can be characterized according to their oligomer number N and compared with the corresponding values for linear aggregate states. These "prenucleation events" may provide some advance measure of the probability that crystals will ultimately form since the appearance of the light scattering spectrum when aggregates of one of the two forms predominate is significantly different. By examining the resultant spectra under a wide variety of conditions of precipitant concentration, pH, temperataure or other factors, there is some possibility that accurate prediction of crystallization success can be achieved. This might abbreviate the long waiting periods generally necessary to evaluate the success or failure of crystallization experiments and help guide the investigator more rapidly to the most promising conditions for achieving nucleation. Experiments to refine and implement these techniques and better evaluate their validity and practicality are currently in progress in several laboratories.[7]

The growth phases of protein crystals may be of relatively short duration, a few hours, or they may be quite prolonged, extending over several weeks. The view is widely held that protein crystal growth is always very slow and necessarily requires long time periods for completion. This impression has undoubtedly grown up because of the traditional procedures employed and their accompanying lore. We have in the past always "set up" the crystallization trials and, loath to disturb them, examined the results only occasionally over long periods. It is frequently only after long time periods that crystals are observed, hence the impression of slow growth. While growth is sometimes indeed slow, the truth is that no one has carefully measured it for a broad range of crystallization conditions or proteins. The first such experiments are only now underway.[79,80] The likelihood is that long time periods are not really essential for growth, but for the formation of stable nuclei, and there is not necessarily a simple correlation between the probability of stable nuclei formation and the subsequent rate of crystal growth.

What is certain, however, is that the rate of crystal growth increases with the degree of supersaturation. The rate of growth is a direct function of the rate at which molecules of protein appear at the growing crystal faces and the rate at which they attach specifically to the face once presented. Thus, when the system is in the labile region, or near to it, the degree of supersaturation is great and growth proceeds at its most rapid pace. In the metastable region, growth is sustained but at decreasing rates as the system moves away from the labile region toward a saturated solution. At equilibrium, there is no net growth.

One might conclude that the best approach would be to force the system to as high a degree of supersaturation as possible to gain the dual advantages of greatest nuclei probability along with highest rate of growth. While this may be successful in some cases, rapid growth is often associated with defect formation in crystals. Too rapid an incorporation of protein monomers into growing crystals may not allow them sufficient opportunity to order, mosaic blocks of unit cells in the crystal may become misaligned, and impurities may be incorporated into the crystal. This is particularly true for proteins because the lattice forces that fix the molecules into place are insecure and the solvent content great.

In addition to problems of proper incorporation into the growing lattice, rapid growth can produce convective currents at the faces of crystals. These convective currents produce perturbations and dislocations in the crystal lattice due to turbulance and the disruption of continuous transport of monomers to the crystal surfaces.

Defects, such as those seen in Figure 12, and dislocations incorporated into growing crystals produce a variety of deficiencies ranging from poorly ordered crystals that do not diffract X-rays well to twinned, striated, and multiple crystals, none of which are suitable for diffraction analysis. Often, it has been shown, these problems can be minimized and useful crystals obtained simply by finding conditions that reduce the rate of growth, lower the degree of supersaturation during growth, or, alternatively, produce stable nuclei at a lower degree of supersaturation.

It was stated above that no net crystal growth occurs at equilibrium, that is, exactly at saturation. It has, however, long been observed that if a mixture of crystals of various sizes are present and in equilibrium with a saturated solution, the small crystals will sometimes slowly disappear and the largest will increase even more in size. Similarly, there are numerous reports of protein crystals appearing from the midst of amorphous precipitate and ultimately growing to large size as the precipitate disappears. While no net increase in the solid state has occured, there is clearly an energetically favorable redistribution of protein.

This phenomenon makes good sense in thermodynamic terms because a collection of molecules in a single large crystal is at a lower energy state than if it were distributed among numerous small crystals or in a precipitate. This follows from the fact that a molecule at the surface of a crystal has a higher potential energy than one in the interior having all of its bonding needs optimally satisfied. Many minute crystals, having a greater surface to volume ratio than a single large crystal would therefore represent a higher energy state. Thus the large crystal is preferred. Similar arguments apply to the amorphous precipitate as well with the additional condition that molecules forming a precipitate are inherently at a higher energy state.

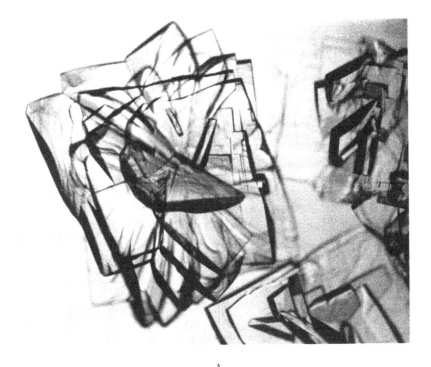

A

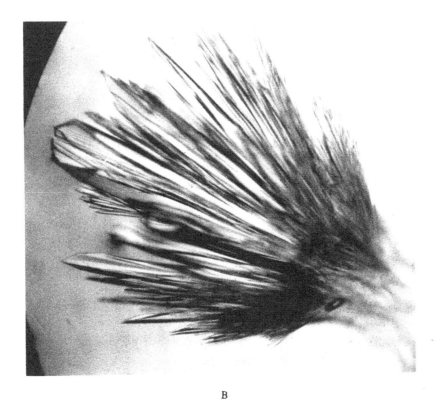

B

FIGURE 12. Examples of crystal twinning and poor growth habits are shown here for orthorhombic crystals of canavalin in (A) and hexagonal prisms of concanavalin B seen in (B). Canavalin is grown from 0.7% NaCl at neutral pH while concanavalin B is grown from 38% saturated ammonium sulfate at pH 6.0.

The willingness of protein crystals to accommodate our needs by readjusting their distribution to larger sizes is a function of the activation energy required to leave the face of one crystal and join the face of another. If this activation barrier is relatively small, the growth of large single crystals may result.

Another phenomenon closely related to that of size redistribution at equilibrium involves polymorphic forms of protein crystals. Examples were shown in Figures 5 and 6. It has occasionaly been observed that a particular habit of crystal will initially form in a sample and after a time a second habit will appear. The first crystal may coexist with the second for a time and then be seen to sowly dissolve as the second grows in size. Ultimately the first crystal disappears completely and the latecomer remains indefinitely. The explanation is that kinetic parameters, dependent on the probability of nuclei formation and the activation energy for monomer accretion, favored the first form. Bonding interactions between molecules in the second lattice, however, were more favorable. The molecules incorporated into the second form were, therefore, at a lower energy state. Thus, kinetics favored the first crystal type, potential energy, the ultimate arbiter, favored the second type. Because the energy differences between polymorphs are small they are often seem to coexist for significant time periods in the same sample. Similarly, very minor changes in crystallization conditions may dramatically shift the equilibrium and thereby completely eliminate one of the forms.

The third phase of crystallization is termed cessation of growth, and it is perhaps the most poorly understood of the three. It derives from the observation that most protein crystals never exceed a fixed maximum size that this size appears to be specific for the particular crystal grown under a fixed set of conditions. There is no obvious reason for this cessation of growth which may in some cases frustrate all efforts to grow crystals suitable for diffraction analysis. It does not appear to occur for crystals of more conventional small molecules.

A number of hypotheses have been suggested to explain why protein crystals seem to reach terminal size and then stop growing. First, it most certainly is not due to depletion of protein monomers from the solution and establishment of equilibrium. This can easily be shown by measurement of solution protein levels in excess of saturation values and could, if it were the cause, be compensated simply by adding new protein solution to the sample. One possibility is that cessation of growth is a consequence of the accumulation of defects in the lattice so that it eventually contains so many errors that the growth faces can no longer propagate. This hypothesis also seems unlikely at this point since there should then be a correlation between crystal perfection as shown by X-ray diffraction and crystal size. Quite the contrary, however, is observed. Some extremely large crystals such as tRNA[43] or α_1-acid glycoprotein[63] hardly diffract X-rays at all while many crystals of small terminal size do so beautifully.

The correct explanation most probably involves surface effects; the contamination of surfaces by denatured protein, by amorphous precipitate, by chemical reactions at the surface or by reduction of physical steps or dislocations that may be essential for continued growth. This is suggested most strongly by the observation that if a crystal of terminal size is cut to expose new faces comprised of formerly internal molecules, the cleaved pieces will then each commence to grow on those faces until both have again reached their terminal size. Similarly, a large seed crystal introduced into a fresh and supersaturated protein solution often will not grow, but if it is first wounded or cut, even slightly, it will proceed to grow. Crystals, no matter what size, whose surfaces have been crosslinked with agents such as glutaraldehyde, never grow.

VIII. THE GENERAL PROPERTIES OF PRECIPITANTS

The most common approach to crystallizing macromolecules, be they proteins or nucleic acids, is to gradually alter the characteristics of a highly concentrated protein solution to achieve a condition of limited supersaturation. As discussed above, this may be achieved by modifying some physical property such as pH or temperature, or through equilibration with precipitating agents. The precipitating agent may be a salt such as ammonium sulfate, an organic solvent such as ethanol, or a highly soluble synthetic polymer such as polyethylene glycol. The three types of precipitants act by slightly different mechanisms, though all share some common properties.

In highly concentrated salt solutions competition for water exists between the salt ions and the polyionic protein molecules. The degree of competition will depend on the surface charge distribution of the protein as well, which is a function primarily of pH. Because protein molecules must bind water to remain solvated, when deprived of sufficient water by ionic competition, they are compelled to associate with other protein moleculees to satisfy their electrostatic requirements. Aggregates may be random in nature leading to linear and branched oligomers, and eventually to precipitate. When the process proceeds in an orderly fashion and specific chemical interactions are used in a repetitive and periodic manner to give three dimensional aggregates, then the nuclei of crystals will form and grow.

The removal of available solvent by addition of precipitant is in principle no different than the crystallization of sea salt from tidal pools as the heat of the sun slowly drives the evaporation of water. It is a form of dehydration but without physical removal of water.

A similar effect may be achieved as well by the slow addition to the mother liquor of certain organic solvents such as ethanol or acetone. The only essential requirement is that at the specific temperature and pH of the experiment, the organic solvent not adversely affect the structure and integrity of the protein. This is often a very stringent requirement and deserves more than a little consideration. The organic solvent competes to some extent like salt for water molecules, but it also reduces the dielectric screening capacity of the intervening solvent. Reduction of the bulk dielectric increases the effective strength of the electrostatic forces that allow one protein molecule to be attracted to another.

Polymers such as polyethylene glycol also serve to dehydrate proteins in solution as do salts, and they alter somewhat the dielectric properties like organic solvents. They produce, however, an additional important effect. PEG perturbs the natural structure of the solvent and creates a more complex network having both water and itself as structural elements. A consequence of this restructuring of solvent is that macromolecules, particularly proteins, tend to be excluded and phase separation is promoted.[32,52]

Crystallizaton of proteins may also be accomplished by increasing the concentration of a precipitating agent to a point just below supersaturation and then adjusting the pH or temperature to reduce the solubility of the protein.

Crystallization using any of the precipitation methods is unpredictable as a rule. Every macromolecule is unique in its physical and chemical properties because every amino acid or nucleotide sequence produces a unique three dimensional structure having distinctive surface characteristics. Thus lessons learned by investigation of one protein are only marginally applicable to others. This is compounded by the behavior of macromolecules which is complex owing to the variety of molecular weights and shapes, polyvalent surface features that change with pH and temperature, and to their dynamic properties.

Because of the intricacy of the interactions between solute and solvent, and the shifting character of the protein, the methods of crystallization must usually be applied over a broad set of conditions with the objective of discovering the particular minimum (or minima) that yield crystals. In practice, one determines the precipitation points of the protein at sequential pH values with a given precipitant, repeats the procedure at different temperatures, and then examines the effects of different precipitating agents.

It is invaluable if before actually initiating crystallization trials, one acquires as much insight into the precipitation behavior of the macromolecule as possible. This may come naturally from experience with the preparation and purification of the molecule, or obtained

TABLE 1
Methods for Attaining a Solubility Minimum

1. Bulk crystallization
2. Batch method in vials
3. Evaporation
4. Bulk dialysis
5. Concentration dialysis
6. Microdialysis
7. Liquid bridge
8. Free interface diffusion
9. Vapor diffusion on plates or slides
10. Vapor diffusion in hanging drops
11. Sequential extraction
12. pH induced crystallization
13. Temperature induced crystallization
14. Crystallization by effector addition

by a series of simple preliminary experiments. The effect of various parameters such as the concentration of precipitating agent may be studied, for example, by observation with a low power microscope of the results of deliberate addition in microliter amounts of precipitant to a microdroplet of mother liquor in the well of a depression slide. This conserves material and narrows the initial range of likely conditions.

IX. METHODS FOR PRODUCING SUPERSATURATION

Most of the methods in common use for inducing supersaturation are listed in Table 1. The most primitive technique used for the crystallization of proteins is to simply add a concentrated salt solution, or some other precipitating agent, directly to the protein solution while observing it visually. When the sample becomes faintly opalescent it is set aside in a still and quiet place. At appropriate intervals of time droplets are removed from the solution and examined under a microscope for the appearance of crystals. Northrup et al.[73,74] and Baranowski[8] used this procedure to crystallize numerous proteins, and the first protein crystals from animal sources, horse serum albumin[58] and ovalbumin,[34] were obtained in this way. It was observed by the early investigators in the field that crystallization was most frequently successful from highly concentrated protein solutions of 10 to 100 mg/ml that had been filtered clear of any contaminants or amorphous material. They further emphasized that the only suitable conditions for crystallization were those that did not denature the protein.

Sumner and Somers,[91,92] two early pioneers, suggested that too high a degree of supersaturation was unfavorable for crystallization and that if a strong opalescence or schlieren effect appeared, it generally implied that an excess of precipitant had been added. They pointed out that a solution saturated with respect to amorphous precipitate is many times supersaturated with respect to crystals.

Once the faint opalescence is observed, however, distilled water may be added back dropwise to the sample until the schlieren just vanishes. The solution can then be placed in the cold and allowed to stand undisturbed. Transfer of the solution from warm to cold temperature mildly increases the solubility of the protein and serves to further lessen the degree of supersaturation. While these procedures seem to work well with salts as the precipitant, when PEG or organic solvents are employed it is frequently difficult to induce the protein to re-enter solution.

There are numerous descriptions in the literature of the application of this technique based on bulk crystallization. Excellent examples were reported by Sumner and Somers,[91,92] Bailey,[3,4] Osborne,[75,76] and Northrop et al.[73,74]

If one knows more or less the conditions under which crystals are likely to form, or even the precipitation point of the protein for a particular precipitating agent then a simple batch method may be employed. In this procedure small glass vials or tubes are filled with 0.5 to 1.0 ml of the 10 to 20 mg/ml protein solution at a precipitant concentration very slightly less than that required to exclude the protein from solution. A range of precipitant concentrations may then be investigated by adding microdrops of saturated precipitant solution with micropipettes to produce a gradient array of samples. The vials are then sealed with caps or stoppers and set aside.

With this, as with most other methods, the difference between amorphous precipitate, microcrystals, and large single crystals may be only a fraction of a percent of saturation by the precipitating agent. Thus, it is essential, that when conditions have been discovered at which some kind of crystals are obtained, that they be carefully refined in small increments and optimized. This applies not only to precipitant concentration, but to pH and other variables as well.

The primary disadvantages of these bulk methods are the relatively large amount of protein solution required and the difficulty in modifying conditions once the vials have been sealed. In addition, they are not well suited for screening a broad range of conditions if crystallization has not previously been affected.

A macromolecule may be guided gradually toward crystallization by dialysis against a solution of salt or organic solvent concentration nearly adequate to produce precipitation. This method has the advantage that as the differential between concentrations on the two sides of the dialysis membrane decreases, the rate of equilibration decreases as well. In addition, dialysis has the advantage that a virtual continuum of precipitant concentrations or range of pH values can be explored by modifying the conditions on the external side of the membrane. If the concentration of precipitant is too high and amorphous rather than crystalline material results, the sample may be redissolved and new conditions established simply by adjusting the external solution.

Dialysis using bulk solutions also requires a considerable amount of material, but it has been adapted to small volume experiments by Zeppenzauer et al.[105,106] and this has yielded admirable success in many laboratories. Various designs have been described for microdialysis cells or vessels.[51,105] In most cases, samples of the protein solution only 10 to 20 μl in volume are injected into short glass capillaries or tubes. These are then sealed at one end and a dialysis membrane is used to cover the other end. The entire assembly is submerged in a test tube or other container holding the precipitant solution. Gradual equilibration then occurs through the membrane.

Weber and Goodkin described one microdialysis apparatus along with a fairly extensive analysis of the diffusion rates as a function of various system variables.[10] This analysis is useful in arriving at suitable alternate designs and facilitates variation of external conditions when a large number of samples are used to screen possible crystallization conditions.

Using the observation that the solubility of most proteins in concentrated ammonium sulfate solutions decrease with an increase in temperature, Jacoby[35,36] devised a technique for protein crystallization based on sequential extraction. With this technique, the protein is first completely precipitated with salt and then collected by centrifugation. The pellet is then alterenately extracted and recentrifuged in a sequential fashion with solutions of serially decreasing concentrations of ammonium sulfate at 4°C until it is fully redissolved. The series of supernatants are then transferred to room temperature thus modestly decreasing the solubility of the proteins in each sample and producing supersaturation. While this procedure has not, in general, proven useful for growing large protein crystals, microcrystals have been produced in numerous instances.

If a protein is soluble in one solvent system and insoluble in another then the solution of least density can be carefully layered atop the other in some kind of glass tube. Transient conditions of supersaturation will be achieved in the region of the interface as the two layers diffuse into one another. Frequently, crystal nuclei will form and grow. This free interface diffusion technique is adaptable to milligram quantities of protein by carrying out the trials in small capillary tubes.[85] Variation of capillary bore even provides a means by which the rate of diffusion can be regulated. The method, furthermore, can be used with salts, organic solvents, or polymeric precipitant systems. The important points to bear in mind are that the protein solution should be highly concentrated, and that flocculant precipitate does not appear at the interface when layering is performed, but only a barely perceptible turbidity.

A modification of the free interface diffusion technique has been employed when layering is a problem. The protein sample is placed in a glass capillary or tube and frozen at $-20°C$. Onto this solid phase is layered the precipitating solution that has been chilled to $0°$ or below. The capillaries are then transferred to 4°C where slow thawing is allowed to occur and finally, if appropriate, to room temperature. For more dense salt solutions, or any precipitant solution more dense than the protein solution, the procedure can be reversed with the protein layered atop the precipitant.

The most interesting proteins, from the standpoint of biochemistry or molecular biology, are frequently those most difficult to obtain in quantities sufficient for bulk crystallization methods. As a consequence, microtechniques have been devised that use in each trial no more than a few microliters of protein solution. An array of microtechniques are shown in Figure 13. With these, several hundred different attempts may be carried out with no more than 20 mg of protein. Principal among these microcrystallization techniques are those based on vapor diffusion.[30,60,61,62]

According to one popular procedure called the "sitting drop", droplets of 10 to 20 μl are placed in the nine wells of depression spot plates (Corning 7220). The samples are then sealed in transparent containers, such as Pyrex dishes or plastic sandwich boxes (Tri-State Plastics, Henderson, Kentucky), which hold, in addition, reservoirs of 20 to 50 ml of the precipitating solution. The plates bearing the protein or nucleic acid samples are held off the bottom of the reservoir by the inverted half of a disposable Petri dish. Through the vapor phase, the concentration of salt or organic solvent in the reservoir equilibrates with that in the sample. The components of this method are illustrated in Figures 14 and 15. In the case of salt precipitation, the droplet of mother liquor must initially contain a level of precipitant lower than the reservoir, and equilibration proceeds by distillation of water out of the droplet and into the reservoir. This holds true for nonvolatile organic solvents, such as PEG or MPD as well. In the case of volatile precipitants, none need be added initially to the microdroplet, as distillation and equilibration proceed in the opposite direction.

The method has the advantage that it requires only small amounts of material and is ideal for screening a large number of conditions. The major disadvantage is that all samples in a single box must be equilibrated against the same reservoir solutions. It does, however, permit considerable flexibility in varying conditions once the samples have been dispensed, by modification of the concentration of precipitants in the reservoir. When clear plastic boxes are used, large numbers of samples can be quickly inspected for crystals under a dissecting microscope and conveniently stored.

Often it is desirable to screen a large number of precipitants or a vast array of precipitant concentrations. The "sitting drop" technique is, therefore, less appropriate than a second method known as the "hanging drop". It too uses vapor phase equilibration but with this approach, a microdroplet of mother liquor (as small as 5 μl) is suspended from the underside of a microscope cover slip, which is then placed over a small well containing 1 ml of the precipitating solution. The wells are most conveniently supplied by disposable plastic tissue culture plates (Linbro model FB-16-24-TC) that have 24 wells (2 cm diameter × 2 cm deep) with flat ground rims that permit airtight sealing by application of silicone vacuum grease around the circumference. These plates provide the further advantages that they can be swiftly and easily examined under a dissecting microscope and they allow compact storage.

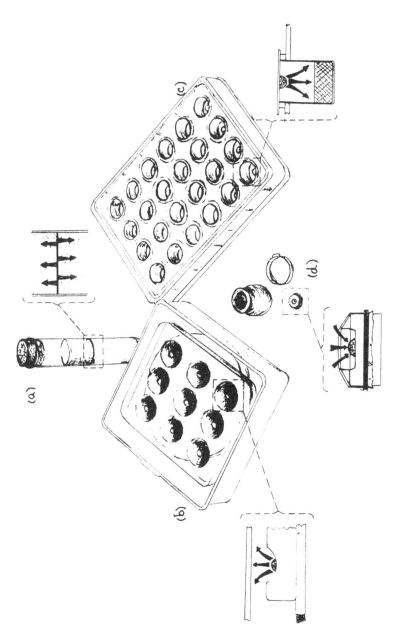

FIGURE 13. *Shown here is an array of the four most common microtechniques currently in use for the crystallization of macromolecules.* (a) The free interface diffusion technique., (b) and (c) are two useful vapor diffusion methods using sitting drops on glass depression plates and hanging drops in tissue culture plates. (d) A liquid dialysis button and a small vial which serves as the exterior liquid reservoir. All can be used with a variety of conditions and precipitating agents and each allows gradual equilibration of the protein and precipitating solutions to attain supersaturation.

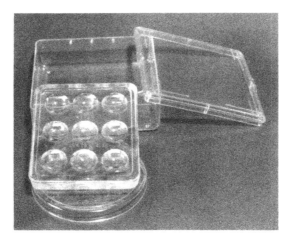

FIGURE 14.   The apparatus for conducting sitting drop vapor diffusion crystallization trials is shown here. It consists of a plastic box which can be sealed from the air and half of a disposable petri dish which serves as a stand for the nine depression glass spot plate in which the protein microdrops are deposited.

The hanging drop technique can be used both for the optimization of conditions and for the growth of large single crystals. A photograph of the hanging drop boxes and cover slips is found in Figure 16.

While the principle of equilibration with both the sitting drop and the hanging drop are essentially the same they frequently do not give the same results even though the reservoir solutions and protein solutions are identical. Presumably because of the differences in the apparatus used to achieve equilibration, the path to equilibrium is different in the two cases even though the end point is the same. In some cases there are striking differences in the degree of reproducibility, final crystal size, morphology, required time, or degree of twinning. These observations illustrate again the important point that the pathway leading to supersaturation may be as important as the final point reached.

A more direct technique that may also be used with microliter amounts of protein sample is the liquid bridge carried out on single depression microscope slides. With this technique, a microdrop of mother liquor and a drop of the precipitating solution are placed in close proximity in the depression of a glass slide. A fine needle is used to draw a thin liquid bridge connecting the two drops so that free diffusion can occur between them. The slide and droplets are then sealed from the air to prevent evaporation. By direct liquid diffusion across the bridge and into the mother liquor, the precipitant induces crystallization.

Some macromolecules will spontaneously crystallize when highly concentrateed at low ionic strength. Thus, numerous initial observations of a protein's crystallization have been made by NMR specialists preparing samples for analysis. Concentration can be accomplished by a number of techniques, but the most practical is to use vacuum dialysis in conical collodian membranes. An ideal apparatus for this purpose is made by Schleicher and Schüll Co. which can, over the course of several hours, concentrate a sample of 20 to 50 ml to a volume of only a fraction of a milliliter. The device has the advantage that the appearance of crystals can be visually detected and their recovery is fairly easy. It is particularly useful because dialysis continues during the concentration process, thus conditions of pH, ionic strength, or effector concentration can be adjusted simultaneously with volume reduction.

As noted earlier, one of the most powerful techniques for producing a supersaturated protein solution is the adjustment of the pH to values where the protein is substantially less soluble. This may be done in the presence of a variety of precipitants so that a large number of possibilities can be created whereby crystals might form. The gradual alteration of pH is

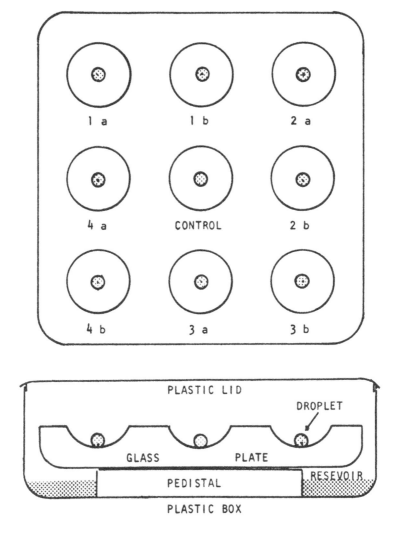

FIGURE 15. A schematic diagram of the sitting drop vapor diffusion apparatus employing the nine well glass depression plate in a plastic box.

particularly useful because it may be accomplished by a variety of gentle approaches that do not otherwise perturb the system or introduce unwanted effects.

Although microdialysis is probably equally suitable, more success has been achieved with the vapor diffusion method using microdroplets on spot plates. The ambient salt, effector, or buffer conditions are establisehd prior to dispensing the microdroplets in the depressions on the plate. The pH is then slowly raised or lowered by adding a small amount of volatile acid or base to the reservoir. Diffusion of the acid or base then occurs from reservoir to sample, just as for a volatile precipitant.

If the pH is to be raised, for example, a small drop of concentrated ammonium hydroxide can be added to the reservoir; a drop of acetic acid may be used to lower it. The pH can also be gradually lowered over a period of days by simply placing a tiny chip of dry ice in the reservoir. The liberated CO_2 diffuses and dissolves in the mother liquor to form weak carbonic acid.

When a specific pH end point is required, the mother liquor can be buffered with suitable compounds at that point and then moved significantly away by addition of acid or base. The microdoplets of mother liquor may then be returned to the buffer point by addition of an appropriate volatile acid or base to the reservoir.

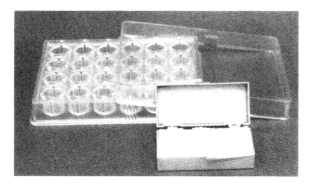

FIGURE 16. A popular arrangement for conducting large numbers of vapor diffusion trials is a 24 well tissue culture tray of clear plastic (Linbro FB-16-24-TC) that provides discrete reservoirs over which can be placed microdrops of mother liquor hanging suspended from the undersides of siliconized circular 22 mm cover slips.

As with pH, proteins may vary in solubility as a function of temperature, and some are very sensitive. One can take advantage of this property with both bulk and microtechniques, although the latter are somewhat less appropriate in this instance. Many of the earliest examples of protein crystallization were based on the formation of concentrated solutions at elevated temperatures followed by slow cooling. Osborne in 1892 reported the crystallization of more than 10 plant seed globulins by cooling relatively crude extracts from 60°C to room temperature in the presence of varying concentrations of sodium chloride.[75] These same procedures were followed by Bailey in 1942[3,4] and Vickery et al. in 1941[98] to crystallize other proteins of the same type. A more recent example is that of glucagon[44] which is crystallized by dissolving the protein at 60°C in appropriate buffers and cooling slowly to room temperature. The crystallizations of insulin[6] and proinsulin,[23] alkaline phosphatase,[31] deoxyribonuclease,[49] and α-amylase[65] also depend at least in part on the cooling of a protein solution.

If temperature change is an important consideration or the primary means for inducing crystal formation, its rate may be manipulated to some extent by enclosing the sample at elevated temperature in a Dewar flask or insulated container and then placing the container at the desired final temperature. The use of thermal insulation in this regard has been reported for insulin[6] and has been used as well for the crystallization of numerous conventional small molecules of biological interest.

The use of temperature is usually of value when the protein solution is at low ionic strength, but this is not always the case. Histidinol dehydrogenase[103] is crystallized by cooling in the presence of salt (apparently contrary to the general rule pointed out by Jacoby[36]). Temperature may also be used in some cases to alter the crystal habit once crystallization conditions are relatively well established. Levansucrease provides an example of this.[9]

X. SEEDING

Often one is attempting to reproduce crystals of a protein previously grown where either the formation of nuclei is limiting, or spontaneous nucleation occurs at such a profound level of supersaturation that poor growth patterns result. In such cases it is desirable to induce growth in a directed fashion at low levels of supersaturation. This can sometimes be accomplished by seeding a metastably supersaturated protein solution with crystals from earlier trials. The seeding techniques fall into two categories, those employing microcrystals as seeds and those using larger macroseeds. In both methods, the fresh solution to be seeded

FIGURE 17. Frequent problems in protein crystal growth are the formation of too many nuclei and growth in habits unsuitable for diffraction analysis. Both difficulties are exhibited here by masses of pig pancreas α-amylase crystals which grow in a thin plate morphology of space group $P2_12_12$.

should be only slightly supersaturated so that controlled, slow growth will occur. The two approaches have been described in some detail by Fitzgerald[22] and by Thaller et al.,[93,94] respectively.

In the method of seeding with microcrystals, the danger is that too many small crystals will be introduced into the fresh supersaturated solution and masses of crystals, like that shown in Figure 17, will result, none of which are suitable for diffraction analysis. To overcome this, a stock solution of microcrystals is serially diluted over a very broad range. Some dilution sample in the series will, on average, have no more than one microseed per microliter. Others will have tenfold more or none. A microliter of each sample in the series is then added to a fresh protein crystallization trial under what is perceived to be optimal conditions for growth to occur. This empirical test should, ideally, identify the correct sample to use for seeding by yielding only one or a small number of single crystals when crystal growth is completed. Seeding solutions having too many seeds will yield showers of other microcrystals and seeding solutions containing too low a concentration of seeds will produce nothing at all. The optimal seeding concentration as determined by the test can then be used to seed many additional samples.

The second approach to seeding involves crystals large enough to be manipulated and transferred under a microscope. Again the most important consideration is to eliminate spurious nucleation by transfer of too many seeds. Even if a single large crystal is employed, microcrystals adhering to its surface may be carried across to the fresh solution. To avoid this, it is recommended that the macroseed be thoroughly washed by passing it through a series of intermediate transfer solutions.[93] In so doing, not only are microcrystals removed, but if the wash solutions are chosen properly, some limited dissolution of the seed may take place. This has the effect of freshening the seed crystal surfaces and contributing to new growth once it is introduced into the fresh protein solution. Again, the new solution must be at least saturated with respect to protein but not extremely so in order to insure slow and proper growth.

Even with these precautions macroseeding still often results in uncontrolled crystallization from secondary nuclei. One additional measure that can be taken is to introduce the seed crystal just at the very edge of the new sample of solution so that there is minimal contact between seed and solution. A short liquid bridge to the bulk of the sample might also be employed. Sometimes, even when there are multiple seeds, a focus of growth at one edge of the sample will predominate and only a few larger crystals will result.

Seeding is frequently a useful technique for promoting nucleation of protein crystals, or initiating nucleation and growth at a lower level of supersaturation than might otherwise spontaneously occur. This can only be done, however, where crystals, even poor crystals, of the protein under investigation have previously been obtained and can be manipulated to serve as seeds. A very common problem in macromolecular crystallization is inducing crystals to grow that have never previously been observed. This reflects, of course, the salient fact that the formation of stable nuclei of protein crystals is most often the single major obstacle to obtaining any crystals at all. Thus, in those cases where the real problem is simply getting crystals, any crystals, attention should be focused on the nucleation problem, and any techniques that might help promote nucleation should be employed.

One such technique, borrowed in part from classical small molecule crystal growth methodology, is the use of heterogeneous or epitaxial nucleants. In principle, this means that induction of growth of crystals of one substance on crystal faces of another. The classical example is galium arsenide crystals that nucleate and grow from the faces of crystals of silicon.

Because protein molecules possess chemical groups, both charged and neutral, that often readily interact with small molecules, membranes, or other surfaces, the possibility presented itself that the faces of natural and synthetic minerals might help order protein molecules at their surfaces and thereby induce the formation of ordered two dimensional arrays of the macromolecules. This ordering might occur by mechanical means due to steps on the crystal faces or by chemical means derived from a complementarity between groups on the mineral and the protein. Such cooperation between mineral crystal face and nascent protein crystal face might be particularly likely should the lattice dimensions of the protein unit cell be integral multiples of natural spacings in the mineral crystal.

Recently, McPherson and Schlicta have shown in a series of experiments using fifty different water insoluble minerals and five different proteins that both heterogeneous nucleation and epitaxial growth of protein crystals from mineral faces do indeed occur.[67] For each of the five proteins, certain specific sets of minerals were empirically identified that promoted nucleation and growth at earlier times and lower levels of supersaturation than occurred through spontaneous events.

Particularly striking was the growth of the protein lysozyme on the substrate apophylite

FIGURE 18. A crystal of lysozyme that has nucleated and grown from the surface of a crystal of the mineral apopholite apparently by an epitaxial mechanism involving a matching of crystal lattices.

shown in Figure 18. As is apparent in the photograph, the protein and mineral crystal faces are parallel and morphologically suggest alignment. A search of the lattice spacings of all mineral crystals and protein crystals utilized in the experiment showed that the 100 face of the tetragonal lysozyme crystal matched a 3 × 5 array of the 110 plane of apophylite with extremely small linear mismatches (0.13 and 0.53%) and an areal mismatch of only 0.4%. Analysis of the crystal-crystal pair by X-ray diffraction photography showed that the lattices of the lysozyme and apophylite crystals were indeed aligned. Thus it was demonstrated that true epitaxial nucleation and growth of protein crystals from the faces of appropriate mineral substrate crystals can occur. The use of such hetero or epitaxial substrates might provide an important mechanism for enhancing or promoting protein crystal growth in cases where nucleation is the principal problem.

A second approach to enhancing the formation of crystal nuclei was described by W. J. Ray.[82] He introduced microdroplets of various concentrations of PEG into protein solutions that were also sufficiently high in salt concentration (approximately 50% saturated with ammonium sulfate) to support crystal growth once stable nuclei were formed. He was able to show that protein left the salt dominated phase of the mixture and concentrated itseslf in the PEG rich microdroplets, sometimes reaching effective concentrations in the PEG droplets of several hundred milligrams per millilter. By light microscopy techniques it was dem-

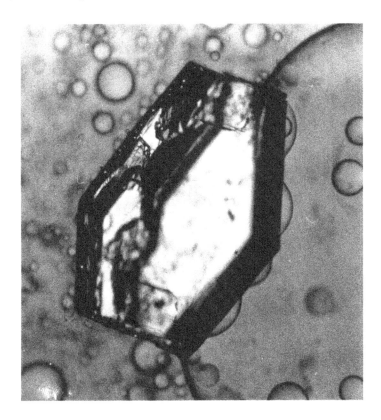

FIGURE 19.   Lysozyme was crystallized from 5% NaCl in the presence of detergent. The concentration of surfactant was sufficiently high that it formed a separate phase as droplets in the lysozyme-salt solutions. Shown here is the growth of a large tetragonal lysozyme crystal that nucleated and grew from the surface of the detergent rich droplet.

onstrated that crystal nuclei appeared first at the surface of the PEG droplets and then proceeded to grow into the superesaturated salt solution that surrounded them, finally reaching a terminal size appropriate for X-ray analysis. In the absence of the PEG droplets no crystals were ever observed to form. A similar effect has also been observed with some detergents in PEG precipitating systems as seen in Figure 19.

These experiments are very encouraging that other, perhaps even more effective, heterogeneous precipitant-solvent systems might be found that will assist in the enhancement of crystal nucleation. Ray refers to the PEG microdroplets as "crystallization catalysts". One would think that further investigation of this phenomenon might be profitable.

XI. VARIABLES IMPORTANT TO CRYSTAL GROWTH

Most of the common methods for the crystallization of proteins rely in one way or another on dehydration of the protein by either removal of solvent, a lessening of the chemical activity of water, or by competition for water. Thus, simple evaporation might seem an appropriate technique, and indeed it has been used successfully.[13] There are many designs of apparatus for producing slow evaporation using wicks or capillaries and the methods of execution of this technique is really a function of the investigator's imagination. The principal disadvantage of evaporation is that it is difficult to carry out in a controlled and reproducible manner. Suddath (personal communication), however, has pointed out that equilbration of protein samples containing precipitant against an inert gas at specified humidity would accomplish the same purpose as the vapor equilibration procedures described above. He has described a closed chamber into which gas can be allowed at specified temperatures and humidities and shown it to be a viable means of producing crystal growth. By including in the chamber sensors for pH, temperature and humidity, and interfacing these to a personal computer, a complete record may be maintained of all parameters during a set of crystallization trials. Alterations in the variables can be accomplished at any time by changing the environment in the chamber, and reproducibility may be attained by continuous monitoring and recording. It appears that further development of this system might indeed be valuable in systematizing the search for growth conditions and an enhanced ability to repeat results.

As emphasized earlier, there are really two approaches to inducing a supersaturated condition and using the nomenclature of Hoffmeister[33] and these are respectively by K_s and β types of effects. β precipitation is usually considered to depend on a change of pH or temperature at a constant salt, or precipitant concentration. pH has perhaps been most frequently used for altering protein solubility, but temperature may also be very useful. The effect of temperature on solubility depends not only on the nature of the precipitant or ionic strength but even more on the particular properties of the protein. Proteins vary greatly in this regard, some being virtually insensitive to temperature change and others, like α-amylase, demonstrating dramatic effects. Temperature may also provide means for modifying the crystal habit once crystallization conditions have been relatively well established.

Most protein and nucleic acids are conformationally flexible or exist in several conformational equilibrium states. In addition, they may assume a substantially different conformation when they have bound coenzymes, substrates, or other ligands.[53] Frequently a protein with bound effector may exhibit appreciably different solubility properties than the native protein. In addition, if many conformational states are available, the presence of effector may be used to select for only one of these, thereby engendering a degree of conformity of structure and system microhomogeneity that would otherwise be absent.

The effect of ligands can be employed to induce supersaturation and crystallization in those cases where its binding to the protein produces solubility differences under a given set of ambient conditions. The effector may be slowly and gently combined with the protein, preferably by dialysis, so that the resulting complex is at a supersaturating level.

The addition of ligands, substrates, and other small molecules has seen widespread use in protein crystallography, since it provides useful alternatives if the apoenzyme itself cannot be crystallized. Lactate dehydrogenase[29], *Staphylococcus* nuclease,[2] dihydrofolate reductase,[48] hexokinase,[90] and many others relied on effector addition, to some extent at least, in the growth of some particular crystal form.

Just as "salting out" may be used to induce crystallization, the opposite effect "salting in" may also be employed. Here the objective is to remove ions and subject the protein to an environment of very low ionic strength. This is done most easily and efficiently simply by dialyzing the protein, in bulk or in micro amounts, against redistilled water. The course of equilibration here may be conveniently monitored with a conductivity meter. Generally crystals produced by this procedure are small or are microcrystals, but this is not always the case.

Dialysis versus distilled water is of course useful for crystallization only when it does not induce irreversible aggregation or denaturation of the protein. The length of time required may sometimes be quite lengthy, several days in some cases, and is best conducted at 4°C. For optimal chance of success the protein concentration should be as high as possible. If microbial contamination is a problem, a drop of toluene or chloroform will suffice.

XII. PRECIPITANTS

Protein precipitants fall into four broad categories: (1) salts, (2) organic solvents, (3) long chain polymers, and (4) low molecular weight polymers and nonvolatile organic compounds. The first two classes are typified by ammonium sulfate and ethylalcohol respectively, and higher polymers such as polyethylene glycol are characteristic of the third. In the fourth category we might place compounds such as MPD and low molecular weight PEG.

As already described, salts exert their effect by dehydrating proteins through competition for water molecules. As predicted by the Hofmeister equation, their ability to do this is proportional to the square of the valences of the ionic species composing the salt. Thus multivalent ions, particularly anions are the most efficient precipitants. Sulfates, phosphates and citrates have traditionally been employed with wide success. A particularly detailed, extensive, and very instructive description of fractionation and crystallization using ammonium sulfate on the proteins from rabbit muscle is given by Czok and Bücher.[19]

One might think that there would be little variation between different salts so long as their ionic valences are the same, or that there would not be much variation with two different sulfates such as Li_2SO_4 and $(NH_4)_2SO_4$. This, however, is often not the case. In addition to salting out, which is a general dehydration effect or lowering of the chemical activity of water, there are also specific protein-ion interactions that may produce other consequences. This is particularly true because of the polyvalent character of the individual proteins, their structural complexity, and the intimate dependence of their physical properties on environmental conditions and interacting molecules.

For example, Li_2SO_4 is a more gentle and often more effective salt for protein crystallization than is the ammonium salt. It is never sufficient, therefore, when attempting to crystallize a protein to only one or two salts and ignore a broader range. Changes in salt can sometimes produce crystals of varied quality, morphology, and in some cases diffraction properties.

It is usually not possible to predict the degree of saturation or molarity of a salt required for the crystallization of a particular protein without some prior knowledge of its behavior. In general, however, it is a concentration just a few percent less than that which yields an amorphous precipitate, and this can be determined for a macromolecule under a given set of conditions using only minute amounts of material.

To determine approximately the precipitation point with a particular agent, a 10 μl droplet of the protein solution can be placed in the well of a depression slide and observed under a low-power light microscope as increasing amounts of saturated salt solution or organic solvent (in 1- or 1-μl increments) are added. If the well is sealed with a cover slip, the additions can be made over a period of many hours, and indeed the droplet should be allowed to equilibrate for 10 or 15 min after each addition, and longer in the neighborhood of the precipitation point. With larger amounts of material the sample may be dialyzed in standard $^1/_4$-in celluloid tubing against a salt solution that is incremented over a period of time until precipitation occurs.

There are rather few salts used in protein crystallization if one surveys the literature and it might be of considerable service if someone were to examine a much larger range of possibilities using a standard set of the more easily crystallized proteins. Undoubtedly salts other than those now in common use wuld be found valuable. A listing of those currently in use is found in Table 2.

As with the salts, few investigations have been conducted with the objective of discovering new organic solvents of value for protein crystalliztion. There are some exceptions, and these are found in References 45 and 46. In general the most common solvents utilized, listed in Table 3, have been ethanol, acetone, butanols and a few other common laboratory reagents.[61] It might be noted here that organic solvents have been of even more general use for the crystallization of nucleic acids, particularly tRNA[43] and the duplex oligonucleotides.[39] Here they have been the primary means for crystal growth. This in part stems from polynucleotides greater tolerance to organic solvents and their polyanionic surfaces which appear to be even more sensitive to dielectric effects than are proteins.

TABLE 2
Salts Used in Crystallization of Proteins

1. Ammonium or sodium sulfate
2. Lithium sulfate
3. Lithium chloride
4. Sodium or ammonium citrate
5. Sodium or potassium phosphate
6. Sodium or potassium or ammonium chloride
7. Sodium or ammonium acetate
8. Magnesium sulfate
9. Cetyltrimethyl ammonium salts
10. Calcium chloride
11. Ammonium nitrate
12. Sodium formate

TABLE 3
Organic Solvents Used in Crystallization of Proteins

1. Ethanol
2. Isopropanol
3. 1,3-Propanediol
4. 2-Methyl-2,4-pentanediol (MPD)
5. Dioxane
6. Acetone
7. Butanol
8. Acetonitrile
9. Dimethyl sulfoxide
10. 2,5-Hexanediol
11. Methanol
12. 1,3-Butyrolactone
13. n-Propanol
14. Pyridine

The only general rules are that organic solvents should be used at a low temperature, at or below 0°C, and they should be added very slowly and with good mixing. Since most are volatile, vapor diffusion techniques are particularly suitable for either bulk or micro amounts. Ionic strength should be maintained as low as possible and whatever means are available should be taken to protect against denaturation.

Reports exist, particularly in the early literature, of protein crystallization from mixed salt-organic solvent systems. Seed globulins were crystallized by Osborne[75], Vickery, et al,[98] and later by Bailey[3,4] from ethanol-NaCl mixtures, for example. There are also several instances of crystallization from high concentrations of salt but in the presence of low amounts of organic solvent such as ethanol.[27,47,71,72] The effect of the small amount of organic solvent in these cases is probably due to a slight reduction in the degree of supersaturation at the equilibrium salt concentration which would promote slower, more orderly crystal growth. Again, much remains to be investigated regarding mixed systems as tools for protein crystal growth.

Polyethylene glycol (PEG) is a polymer produced in various lengths, from several to many hundred units. PEG significantly perturbs the natural structure of water and replaces it with a more complex network having both water and itself as structural elements. One result of this restructuring of the solvent is that macromolecules, particularly proteins, tend to be excluded and phase separation is promoted. In addition to its volume exclusion property, it probably shares some characteristics with salts that compete for water and produce dehydration, and with organic solvents which modify the dielectric properties of the medium. Several studies of the interaction between protein and PEG molecules has appeared that examine the thermodynamic properties of such systems.[32,52]

Aside from its general applicability and utility in obtaining crystals for diffraction analysis,[59] it also has the advantage that it is most effective at minimal ionic strength and it provides a low electron density medium. The first feature is important because it provides for higher ligand binding affinities than does a high ionic strength medium such as concentrated salt. As a consequence there is greater ease in obtaining isomorphous heavy-atom derivatives and in forming protein-ligand complexes for study by difference Fourier techniques. The second characteristic, a low electron-dense medium, implies a generally lower background or noise level for protein structures derived by X-ray diffraction and presumably, therefore, a more ready interpretation.

A number of protein structures have now been solved using crystals grown from PEG. These along with several studies of a more preliminary nature confirm that the protein molecules are in as native a condition in this medium as in those traditionally used. This is perhaps even more so, since the larger molecular weight PEGs probably do not even enter the crystals and therefore do not directly contact the interior molecules. In addition, it would seem that crystals of a specific protein when grown from PEG are essentially isomorphous with, and exhibit these same unit cell symmetry and dimension as, those grown by conventional means.

Polyethylene glycol is produced in a variety of polymer sizes. The low molecular weight species are oily liquids while those above 1000 at room temperature exist as either waxy solids or powders. The latter are preferable for easy dissolution. The size specified by the manufacturer is the mean molecular weight of the polymeric molecules, and the distribution of weights about that mean may vary appreciably. It is certainly broad for the very high molecular weight species. The most popular sizes currently in use are 1000, 4000, 6000, and 20,000. PEG in its commercial form does contain contaminants; this is particularly true of the high molecular weight forms such as 15,000 or 20,000. These may be removed by simple purification procedures[27] or in the case of PEG 20,000 by dialysis in low-pass dialysis or collodian tubes. There have been reports that repurified PEG has proven more effective, and certainly the continaminants could be disadvantageous for some proteins.[38]

All of the PEG sizes from 400 to 20,000 have successfully provided protein crystals, but the most useful are those in the range 2000 to 6000. There have appeared a number of cases, however, in which a protein could not be easily crystallized using this range but yielded crystals in the presence of 400 or 20,000. The molecular weight sizes are generally not completely interchangeable for a given protein even within the mid range, some producing the best formed and largest crystals only at, say, 4000 and less perfect examples at other weights. This is a parameter which is best optimized by empirical means along with concentration and temperature.

A very distinct advantage of polyethylene glycol over other agents is that most proteins (but not all) crystallize within a fairly narrow range of PEG concentration; this being from about 4 to 18%. In addition, the exact PEG concentration at which crystals form is rather insensitive and if one is within 2 or 3% of the optimal value some success will be achieved. With most crystallizations from high ionic strength solutions or from organic solvents, one must be within 1 or 2% of an optimum lying anywhere between 15 and 85% saturation. The great advantage of PEG is that when conducting a series of initial trials to determine what conditions will give crystals, one can use a fairly coarse selection of concentrations and over a rather narrow total range. This means fewer trials with a corresponding reduction in the amount of protein expended. Thus it is well suited for particularly precious proteins of very limited availability.

The time required for crystal growth with PEG as the precipitant is also generally much shorter than with ammonium sulfate or MPD but occasionally longer than required by volatile organic solvents such as ethanol. Although equilibration times will depend on the differential between starting and target concentrations, if this is no more than 3 or 4%, then crystallization may occur within a few hours or a few days. It seldom requires more than 3 weeks. Thus evaluation of results can be made without undue demands on patience. It should be noted that protein-PEG solutions are excellent media on which to grow microbes, particularly molds, and if crystallizatin is being attempted at room temperature or over extended periods of time, then some retardant such as azide (commonly 0.1%) must be included in the protein solutions.

Since PEG solutions are not volatile, PEG must be used like salt and equilibrated with the protein by dialysis, slow mixing, or vapor equilibration. This latter procedure, utilizing either 10 μl hanging drops over 0.5 ml reservoirs or 20 μl drops on multidepression glass plates in a sealed chamber, has proven the most popular. The author has found that when the reservoir concentration is in the range of 5 to 12%, the protein solution to be equilibrated should be at an initial concentration of about half. That is conveniently obtained by adding 10 μl of the reservoir to 10 μl of the protein solution. When the final PEG concentration to be obtained is much higher than 12%, it is advisable to start the protein equilibrating at no more than 4 to 5% below the final value. This reduces unnecessary time lags during which the protein might denature.

Crystallization of proteins with PEG has proven more successful when the ionic strength is low. It is quite difficult when ionic strength is high. Good buffer conditions in the neutral range are, for example, 10 to 40 mM Tris or cacodylate buffer. If crystallization proceeds too rapidly, addition of some neutral salt may be used to slow growth and better effect crystal form. PEGs are useful over the entire pH range and over a broad temperature range and show no anomalous effects in repsonse to either. PEG appears to be an excellent crystallization agent over the whole spectrum of proteins, although in specific cases other precipitants may be superior.

XIII. FACTORS INFLUENCING PROTEIN CRYSTAL GROWTH

Table 4 lists physical, chemical, and biological variables that may influence to a greater or lesser extent the crystallization of proteins. The difficulty in properly arriving at a just assignment of importance for each factor is substantial for several reasons. Every protein is different in its properties and, surprisingly perhaps, this applies even to proteins that differ by no more than one or just a few amino acids. There are even cases where the identical protein prepared by different procedures or at different times may show significant variations. In addition, each factor may differ considerably in importance for individual proteins. α-Amylase, for example, is strikingly sensitive to temperature change, while concanavalins A and B show little or no variation in crystallization properties as a function of temperature.

Because each protein is unique, there are few means available to predict in advance what specific values of a variable, or sets of conditions might be most profitably explored. Finally, the various parameters under one's control are not independent of one another and their interrelations may be complex and difficult to discern. It is, therefore, difficult to elaborate a rational set of guidelines relating to physical factors or ingredients in the mother liquor that can increase the probability of success in crystallizing a particular protein. The specific components and conditions must be painstakingly sought and refined for each macromolecule.

As already noted, temperature may be of great importance or it may have little bearing at all. In general, it is wise to initially duplicate all crystallization trials and conduct parallel investigations at 4 and at 25°C. Even if no crystals are observed at either temperature, differences in the precipitation behavior of the protein with different precipitants and with various effector molecules may give some indication as to whether temperature will play an important role. If crystals are observed to grow at one temperature and not, under otherwise identical conditions, at the other, then further refinement of this variable is necessary. This

TABLE 4
Factors Affecting Protein Crystal Growth

1. pH
2. Ionic strength
3. Temperature
4. Concentration of precipitant
5. Concentration of macromolecule
6. Purity of macromolecules
7. Additives, effectors, and ligands
8. Organism source of macromolecule
9. Substrates, coenzymes, inhibitors
10. Reducing or oxidizing environment
11. Metal ions
12. Rate of equilibration
13. Surfactants or detergents
14. Gravity
15. Vibrations and sound
16. Volume of crystallization sample
17. Presence of amorphous material
18. Surfaces of crystallization vessels
19. Proteolysis
20. Contamination by microbes
21. Pressure
22. Electric and magnetic fields
23. Handling by investigator

is accomplisheed by conducting the trials under the previously successful conditions over a range of temperatures centered on the one that initially yielded crystals.

The only general rules with regard to temperature seem to be that proteins in a high salt solution are more soluble in colder than in a warmer temperature. Proteins, however, generally precipitate or crystallize from a lower concentration of PEG, MPD, or organic solvent at cold temperature than at warmer temperature. One must remember, however, that diffusion rates are less and equilibration slower at cold temperature than higher temperature, so that the times for precipitation or crystal formation may be longer in the cold.

After precipitant concentration, the next most important variable in protein crystal growth appears to be pH. This follows logically since the charge character of a protein and all of its attendant physical and chemical consequences are intimately dependent on the ionization state of the amino acids that comprise the macromolecule. Not only does the net charge on the protein change with pH, but the distribution of those charges, the dipole moment of the protein, its conformation, and in many cases its aggregation state. Thus an investigation of the behavior of a specific protein as a function of pH is perhaps the single most essential analysis that should be carried out in attempting to crystallize the macromolecule.

As with temperature, the procedure to follow is to first conduct parallel crystallization trials at course intervals over a broad pH range and then repeat the trials over a finer grid of values in the neighborhoods of those that initially showed promise. The only limitations on the breadth of the initial range screened are the points at which the protein begins to lose activity and show indications of denaturation. In refining the pH for optimal growth, it should be recalled that the difference between amorphous precipitate, microcrystals, and large single crystals may be only a few tenths of a pH unit.[105]

In addition to adjusting pH for the optimization of crystal size, it is sometimes also useful to explore variation of pH as a means of altering the habit or morphology of a crystalline protein. This is occasionally necessary if the initial crystal form is not amenable to analysis because it grows as fine needles or flat, thin plates or demonstrates some other unfavorable tendency such as striation or twinning.

There have been virtually no systematic studies of such factors as pressure, sound, vibrations, electrical and magnetic fields, or viscosity on the rate of growth, or final quality of protein crystals. Similarly, studies are only now being undertaken to evaluate the effects of convection and fluid flow on protein crystal growth, final size, and perfection. Thus it is not possible at this time to evaluate their influence.

Historically, those interested in the question of protein crystal growth, with some exceptions, focused their efforts on obtaining crystals of only one or a few specific proteins for diffraction analyses. These analyses would then consume years or even lifetimes of study. Few investigated protein crystal growth from a phenomenological perspective. This is now beginning to change. As the time required to solve the structure of a protein crystal decreases to months, and conceivably weeks, then the rate limiting step in the discovery and analysis of new protein structures becomes tied to the rate at which new protein crystals can be grown.

Because the objective has shifted away from data collection and analysis to crystal growth, more and more effort has been turned to the systematic study of the parameters that effect it. Apparent however, from the list of unexplored variables noted above, much remains to be done.

XIV. SOME USEFUL CONSIDERATIONS

As noted by the earliest cultivators of protein crystals, the concentration of protein in the mother liquor should be as high as possible, 10 to 100 mg/ml. This is particularly true if one is attempting to grow crystals of a specific protein for the first time. Undoubtedly probabilities of obtaining crystals are enhanced by simply increasing protein concentration. This alone is sometimes sufficient to push the system into a state of supersaturation and into the labile region where stable nuclei can form. This may not, however, be the best approach in growing large, perfect crystals once optimal conditions for all other parameters have been established.

Nuclei formed at very high protein concentrations are generally at a supersaturation level far above the metastable region and, as one would expect, their initial growth is extremely rapid. Attendant to the swift initial growth are both the rapid accumulation of defects and imperfections in the growing crystals, and the appearance of many secondary nuclei. The first of those leads to large, flawed crystals exhibiting twinning, striations, or internal lack of order. The secondary nuclei result in showers, indeed storms, of microcrystals that are too small to be useful in X-ray diffraction analysis.

Thus it follows, that once conditions for nucleation and growth have been established and the screening of variables more or less complete, the concentration of the protein should be gradually reduced in increments to slow the growth of the crystals. As a general rule, the largest and most perfect crystals result when the rate of accretion of molecules is slow and orderly. Reduction of concentration is an effective means for controlling this.

The time required for the appearance and growth of protein crystals is quite variable and may range from a few hours in the best of cases to several months in others. Because no truly systematic investigations have been carried out, how rapidly crystals grow once visible nuclei have formed remains in question. The rate of growth may not at all reflect the total amount of time required to obtain crystals adequate for analysis. This includes the time required for solvent equilibration to be achieved, for crystal nuclei to form, and for full growth to occur. None of these periods of time have been carefully determined for any proteins.

When one is screening variables to establish optimal parameters, then the practical objective is to have crystallization occur at the greatest possible speed to expedite determination of most probable conditions. When optimizing and refining crystallization parameters, then time itself becomes an important parameter and long periods of slow growth may be desirable.

One caution is in order. If it is observed that a long period elapses without the formation of crystals and then, well beyond the time required for solvent equilibration to have occurred, crystals suddenly appear, then some likely causes should be explored. One possibility is that the protein has, over the long time period, undergone some physical or chemical change. It may have undergone limited hydrolysis, lost a coenzyme or metal ion, or undergone a slow conformational change. By forcing this same event to occur before the crystallization trials are carried out the time required for growth may be substantially reduced. Another possibility is that the apparatus in which the crystallization experiments were carried out was leaking and that very slow dehydration occurred. Thus the final concentration of precipitant may have been appreciably higher than believed. A final possibility is change in the ambient temperature. This is particularly likely when crystallization is being carried out at room temperature and heating or air conditioning systems are switched on and off as the seasons change.

The most intriguing questions with regard optimizing crystallization conditions concern what additional components or compounds should comprise the mother liquor in addition to solvent, protein, and precipitating agents. The most probable effectors are those which maintain the protein in a single, homogeneous, and constant state. Reducing agents such as glutathione or β-mercaptoethanol are useful to secure sulfhydryl groups and prevent oxidation. EDTA and EGTA are good if one wishes to protect the protein from heavy metal or transition metal ions or the alkali earths. Inclusion of these components may be particularly desirable when crystallization requires a long period of time to reach completion.

When crystallization is carried out at room temperature in PEG or low ionic strength solutions, then attention must be given to preventing the growth of microbes. These generally secrete proteolytic enzymes that may have serious effects on the integrity of the protein under study. Inclusion of sodium azide or thymol at low levels may be necessary to discourage invasive bacteria and fungi.

Substrates, coenzymes, and inhibitors often serve to fix an enzyme in a more compact and stable form. Thus a greater degree of structural homogeneity may be imparted to a population of macromolecules and a reduced level of dynamic behavior achieved by complexing the protein with a natural ligand before attempting its crystallization.

In some cases an apoprotein and its ligand complexes may be significantly different in their physical behavior and can, in terms of crystallization, be treated as almost entirely separate problems. This may permit a second or third chance of growing crystals if the native apoprotein appears refractile. Thus, it is worthwhile, when determining or searching for crystallization conditions, to explore complexes of the macromolecule with substrates, coenzymes, analogues, and inhibitors. In many ways, such complexes are inherently more interesting in a biochemical sense than the apoprotein when the structure is ultimately determined.

It should be pointed out that just as natural substrates or inhibitors are often useful, they also can have the opposite effect of obstructing crystal formation. In such cases, care must be taken to eliminate them from the mother liquor and from the purified protein before crystallization is attempted. This is exemplified by many sugar binding proteins such as lectins. Concanavalin A and Abrus lectin can only be crystallized with great difficulty or not at all when glucosamine or galactose, respectively, are present. Pig pancreas α-amylase can also be crystallized only after residual oligosaccharides are removed from the preparation.

Finally, it should be noted that the use of inhibitors or other ligands may sometimes be invoked to obtain a crystal form different from that grown from the native protein. When crystals of the apoprotein are poorly suited for analysis, this may provide an alternative approach.

Although it was noted that microbial growth often resulted in proteolysis of protein samples and was to be avoided, this is not always the case. It has been shown in a number of instances[68,88,89,99,102] that limited and controlled proteolytic cleavage of a protein could render it crystallizable when in the native state it was not. In other cases limited proteolysis resulted in a change of crystal form to a more suitable and useful habit.[41] It should be emphasized that these represent examples of controlled proteolysis where the end product is an essentially homogeneous population of molecules.

Proteases occasionally seem to trim off loose ends or degrade macromolecules to large, stable, compact fragments. These abbreviated proteins are, as a result, more invariant, less conformationally flexible and they often form crystals more readily than the native precursor. Although one might prefer the intact protein, a partially degraded form sometimes exhibits the activity and physical properties that are of primary interest. If a molecule can undergo limited digestion, this form should also be considered in the crystallization strategy.

Various metal ions have been observed in some cases to induce or contribute to the crystallization of proteins and nucleic acids.[61] In some instances these ions were essential for activity and it was therefore reasonable to expect that they might aid in maintaining certain structural features of the molecule. In other cases metal ions, particularly divalent metal ions of the transition series, were found that stimulated crystal growth but played no known role in the macromolecules' activity. One of the oldest examples of an animal protein being crystallized is horse spleen ferritin that forms perfect octahedra when a solution containing the protein is exposed to concentrations of Cd^{++} ion.[25] α-Lactalbumin was similarly shown to crystallize in the presence of this ion[26] and several varieties of α-amylase crystallize spontaneously when presented with Ca^{++} ions.[65,66]

Metal ions should be included for investigation in that class of additives that for any reason might tend to stabilize or engender conformity by specific interaction with the macromolecule.

XV. THE IMPORTANCE OF PROTEIN PURITY AND HOMOGENEITY

With regard to the rate of growth of protein crystals, there are two important effects to consider, the transport of molecules to the face of a growing nucleus or crystal, and the frequency with which the molecules orient and attach themselves to the growing surface. Crystal growth rates can thus be considered in terms of transport kinetics and attachment kinetics. For protein crystals which grow relatively slowly, transport kinetics, dependent primarily on physical forces and movements in the solution phase, is almost certainly the least important of the two. There is not much doubt that the predominant limitation on the rate at which protein crystals grow is, at least over most of the period of growth, a function of the rate of attachment.

The capture of molecules by a growing crystal surface requires, as in any multicomponent chemical reaction, that the molecule to be incorporated have the correct orientation when it approaches the surface and second that it be in the proper chemical state to form bonds essential for fixing it to a set of neighbors. Although there may be some things we can do to improve the statistical probability of proper orientation, there is not likely very much. On the other hand, we may have many opportunities to effect the frequency of attachment by enhancing the number and strength of the bonds between molecules in the lattice. We do this, for example, by optimizing the charge state of the proteins by adjusting pH, providing electrostatic crossbridges by adding metal ions such as Ca^{++}, or by minimizing the dielectric shielding between potential bonding partners by adding organic solvents such as ethanol.

Certainly one major means of promoting periodic bond formation is to insure that the population of molecules to be crystallized are as homogeneous as possible. As suggested

TABLE 5
Sources of Microheterogeneity

1. Presence, absence, or variation in a bound prosthetic group, substrate, coenzyme, or metal ion
2. Variation in the length or composition of the carbohydrate moiety on a glycoprotein
3. Proteolytic modification of the protein during the course of isolation or crystallization
4. Oxidation of sulfhydryl groups during isolation
5. Reaction with heavy metal ions during isolation or storage
6. Presence, absence, or variation in post-translational side chain modifications such as methylation, amidination, and phosphorylation
7. Microheterogeneity in the amino or carboxy terminus or modification of termini
8. Variation in the aggregation or oligomer state of the protein association/dissociation
9. Conformational instability due to the dynamic nature of the molecule
10. Microheterogeneity due to the contribution of multiple but nonidentical genes to the coding of the protein, isozymes
11. Partial denaturation of sample
12. Different animals or sources of enzyme preparations

by Table 5, this is not always straightforward. It means not only that contaminating proteins of unwanted species be eliminated, but that within a target population all individuals assume absolute physical and chemical conformity. Because crystals have as their essential elements perfect symmetry and periodic translational relationships between molecules in the lattice, then nonuniform protein units cannot enter the crystal. They will not bear a proper correspondence to their neighbors. Thus, imperfect molecules will serve as inhibitors of crystal growth and bear a generally negative effect on the attachment kinetics. Should they enter the lattice in spite of their nonconformity, they will introduce flaws which, by accumulation, will ultimately produce defects, dislocations, and imperfections in the crystals.

Until the late 1950s the primary reason that crystallization was so prized was that it provided a highly effective means of separating a pure protein from an impure preparation. This was certainly the rationale behind the work of Sumner, Herriott, Northrop, Kunitz, Osborne, and others who developed the early methods. All the crystalline proteins obtained before 1948 were, therefore, invariably grown from crude mixtures of proteins, and in some cases, very crude mixtures. This is illustrated most dramatically by the report of Grannick that ferritin could be crystallized *in situ* simply by placing droplettes of $CdSO_4$ solution directly on thin slices of horse spleen,[25] by Osborne's observation that excelsin could be crystallized *in vivo* by placing droplets of ether on thin slices of Brazil nut,[75] and by the finding that lysozyme could be crystallized directly from egg white simply by adding seed crystals.[1] These reports, if nothing else, demonstrated that a macromolecular preparation need not be of high purity in order to yield crystals, and frequently, crystals of adequate quality for X-ray analysis.

It is also true that each of these workers, as well as later investigators, reported considerable increases in size and improvements in form when recrystallization was carried out. It seems fair to conclude that, with few exceptions, the probability of successfully growing large single crystals is greatly increased by improved homogeneity of the sample. In some cases minor protein contaminants are the primary impediment to large single crystals, or even microcrystals, and when the macromolecule is subjected to additional purification, crystallization then proceeds rapidly.

It is, for proteins difficult to crystallize, essential to take all possible measures to purify the protein free of contaminants and to do whatever is necessary to engender a state of maximum structural and chemical homogeneity. Frequently we are misled by our standard analytical approaches, such as PAGE or IEF, into believing that a specific protein preparation is completely homogeneous. This is often born out by distinctive differences in the crystallizability of several preparations even when the analyses indicate they are the same. These imperceptible differences may be due to various degrees of microheterogeneity within preparations that lie at the margin of our ability to detect them. Table 5 lists a number of possible causes for microheterogeneity. Others could undoubtedly be added.

The pronounced effects of microheterogeneity on protein crystallization have recently received much more attention from investigators. Giege et al. have discussed this point in detail and provided broad evidence that purification plays a crucial role in successful crystal growth.[24] Bott et al. similarly showed the pronounced beneficial effects of isoelectric focusing on otherwise "pure" protein.[10]

There are occasions, however, when even the most intense effort to crystallize a particular protein fail in spite of the best efforts at ultrapurification and elimination of microheterogeneity. When this is the case, discretion becomes the better part of valor, and it is often wise to turn to a different source of the protein. Often only very small variations in amino acid sequence, as found for example between different species of organisms, is enough to produce dramatic differences in the crystallization behavior of a protein. Thus if the protein from one source proves intractable, consider another.

Along similar lines, it might be noted that proteins manufactured in bacteria by recombinant DNA techniques appear to be especially favorable for crystallization. There are numerous reports of such crystalline proteins in the literature. Proteins produced in this way are apparently less subject to post translational modifications and many of the other sources of microheterogeneity that characterize naturally occurring proteins. Because this technique also provides a means of amplifying the available quantity of otherwise scarce or rare proteins, it will undoubtedly play an important role in future protein crystallization strategies.

Finally, with regard the purity of a protein sample, never trust anyone. No matter who the source, analyze the protein sample yourself to insure that it is not so obviously heterogeneous to begin with that your efforts will be hopeless. Be skeptical of claims of purity unless you have seen the evidence yourself.

From discussions concerning crystallization of macromolecules, one invariably hears of the irreproducibility in growing crystals of a particular protein or nucleic acid. Nearly every structural investigation so far conducted has been cursed at some time or another with a crystal drought during which time progress came to a virtual halt. It occasionally seems that sequences of successes and failures occurring in no discernable order are the norm. This is frequently attributed to an inability to precisely duplicate past successful crystallization conditions.

It appears more likely that the real problem lies not in reproducing crystallization conditions but in isolating a protein or nucleic acid preparation that is uniformly the same each time. While the actual dispensing of a macromolecule for crystallization, once conditions are known, is extremely simple, even trivial for the most part, isolation procedures are by their very nature just the opposite. They allow great opportunity for slight variation and change from preparation to preparation. It is a common experience that one preparation may yield excellent crystals over a carelessly wide range of conditions, while other preparations bear no crystals despite intensive effort. This variability is often not detectable by the usual physical-chemical techniques, such as gel electrophoresis, chromatography, and specific activity measurements, and it often remains a mystery.

The author has numerous times attempted to crystallize samples of a protein or nucleic acid isolated from three or four different preparations and indistinguishable by the usual procedures. In some cases the samples occupied the same equilibration chamber or dialysis vessel. The common result was striking variability in crystal formation from one sample to another. The lesson taught by these experiences is that attempts at crystallization should not be limited to the product of a single preparation but should encompass as many different preparations as is practical. The fate of a macromolecule, as regards to crystallization, is in a great many cases, already determined by the time it reaches the crystallographers hands, and there is sometimes little he or she can do in a positive sense to effect its conversion into crystals.

In general it is true that proteins most amenable to crystallization are those that possess a high degree of stability, are geometrically compact, and that have few post-translational modifications. The most difficult are those that are labile to heat, pH, or other mild influences such as ionic strength. Proteins that exhibit a highly dynamic character, that equilibrate among a set of conformers or aggregation states are generally problems. Proteins that possess covalently linked oligosaccharide moieties or a range of amidations, methylations, phosphorylations, or other alterations are also usually difficult to crystallize.

As already noted, enhanced stability or homogeneity can sometimes be achieved by mild digestion with proteases which trim the flexible components, by lowering temperature to obtain a more homogeneous aggregation state, or by changing sources of the protein to one yielding a more suitable preparation. Increased homogeneity can also be attained in some cases by saturating the protein with a coenzyme, an inhibitor, or whatever effector molecules might help lock the protein into a single state.

XVI. AGGREGATION OF MACROMOLECULES: SPECIFIC AND NONSPECIFIC

In some investigations it was found advantageous to include in the mother liquor, flexible, extended, charged molecules. These serve as electrostatic crosslinking agents between charged chemical groups on macromolecules and assist in ordering and fixing them into a lattice. Because the crosslinking molecules are small they can diffuse into and out of the crystal. Because the interactions are not covalent but electrostatic in nature, the links can freely make and break as the orientations of the protein molecules adjust to find an energy minimum, i.e., as they optimize their bonding arrangements with neighbors.

There are at least two good examples in the literature of the use of these electrostatic crosslinking agents in addition to the many cases where divalent ions serve a similar purpose. In the crystallization of both transfer RNA[43] and fragments of DNA[39] small polyamines such as spermine, putracine, cadaverine, and spermidine were essential in obtaining highly ordered samples. As one might have predicted, the polyamines were observed to link by salt bridges different nucleic acid molecules in the lattice. In another series of studies, it was shown that protein-nucleic acid complexes of RNase A (or RNase B) plus $d(pA)_4$, $d(pA)_6$, or $d(pT)_4$ could be crystallized more easily than the protein alone.[11] Solution of the structures of these complexes showed that within the crystals, the protein molecules were electrostatically crosslinked by deoxyoligomers that spanned, with their negatively charged polyphosphate backbones, the positively charged lysine and arginine groups of adjacent RNase molecules. Further research into the use of polyvalent electrostatic crosslinkers of differing lengths and chemical character might prove to be quite profitable for proteins having distinctively charged surfaces.

The influence of mild detergents on the crystallization of membrane proteins is discussed thoroughly and in detail elsewhere in this volume, but it is useful to point out here that detergents may also be of substantial value in the crystallization of otherwise soluble proteins as well. Many protein molecules, particularly when they are highly concentrated and in the presence of precipitating agents like PEG or MPD tend to form transient and sometimes stable nonspecific aggregates. The presence of such an array of varying sizes, shapes, and charges is not appreciably different than trying to crystallize a protein from a hetereogeneous mixture of an impure solution composed of dissimilar macromolecules. An objective in crystallizing proteins is to limit the formation of nonuniform states and reduce the population to a set of standard individuals that can form identical interactions with one another.

Nonspecific aggregation is primarily a consequence of hydrophobic interactions between

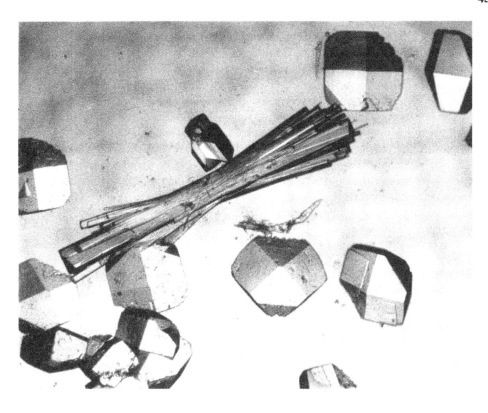

FIGURE 20. Lysozyme under normal crystallization conditions yields only the tetragonal crystal form, but in the presence of some detergents, as shown here, a mixture of the tetragonal and a prismatic habit are obtained. The crystallographic unit cells of the two crystals are distinctly different.

molecules. These place few geometrical constraints on the orientations and bonding patterns between molecules that make up an oligomer. They simply make them stick together in more or less random fashion. Hydrogen bonds and arrays of electrostatic interactions on the other hand generally demand geometrical complementarity between the protein carriers in order to form. They thereby force the macromolecules to orient themselves in specific ways with respect to one another. Thus an objective in obtaining crystals of a protein is to discourage hydrophobic interactions and to encourage those having an electrostatic basis.

A means for limiting nonspecific aggregation is the inclusion of mild, usually nonionic, detergents in the crystallization mother liquor. McPherson, et al. have shown that for a fairly wide range of proteins the neutral detergent β-octylglucoside was a positive factor in obtaining crystals useful for diffraction analysis.[64] In addition, it was demonstrated that other detergents also exhibit helpful properties in altering crystal morphology (an example is shown in Figure 20), decreasing microcrystal formation, or improving growth patterns. One surprising finding that came from these experiments was the high concentration of detergent that most proteins could tolereate without loss of their ability to crystallize and, therefore, presumably their native structure as well. Again, additional research in this area is sorely needed to find those detergents that could yield real advantage.

XVII. ADDITIONAL CONSIDERATIONS AND MEASURES

Often the success of a grand enterprise depends on attention to the smallest detail, and this is true in protein crystallization as well. Care must be excercised at every turn. When attempts are being made to grow large, single crystals a number of added precautions are advised.

1. Crystallization glassware, such as vials, dialysis cells, and diffusion plates, can be surfaced with a silicone coating to make them smooth and non-wetting. This reduces in some cases the number of nucleation points on the glass and tends to result in larger crystals. If insufficient nucleation is a problem, then it may be controlled to some extent by scratching the surface after coating or by adding heterogeneous nucleants.

2. It is important that crystallization samples be maintained free of microbial growth. Conditions of high salt or the presence of alcohol solutions are usually sufficient to preclude such growth, but protein solutions at low ionic strength are ideal for bacteria and fungi. To assure sterility, the macromolecular solution and all buffers may be conveniently drawn through a millipore filter or similar ultramembrane. In addition, this eliminates any amorphous precipitate or debris that could interfere with crystal growth. Very small amounts of toluene, chloroform, or pyridine can also be added to the sample and these are quite effective in preventing microbial growth.

3. It is usually advisable, and often essential, to remove any amorphous material from the mother liquor before it is set up for crystallization. This is best done by centrifuging the sample at high speed immediately before dispensing it. This serves as well to eliminate dust particles and other interfering material. In some cases it may also help remove large macromolecular aggregates that have formed.

4. To further eliminate dust and other contaminants, depression plates, tubes, vials, and other vessels may be sprayed with clean pressurized air or some inert gas to blow away dust just prior to dispensing the mother liquor.

5. Larger volumes of mother liquor usually result in the growth of larger crystals. This is in part due to the inherently greater physical stability of larger volumes. Large volumes act essentially as physical buffers in that they are less subject to vibrations, impact, and thermal fluctuations. Once the optimum conditions for crystallization have been established on a microscale, it is sometimes advantageous to increase sample sizes to at least 40 μl and larger if the amount of material permits.

6. Care should be taken to insure that, if the crystallization glassware has been soaked in acid, it is adequately rinsed or desoaked. This is particularly true of the common laboratory chromic-sulfuric acid cleaning solution, which is not easily rinsed off.

7. If crystallization is to be carried out at low temperature, care should be taken to use buffers that remain soluble. Phosphate buffers are a particular problem in this regard.

8. A great many protein and nucleic acid crystals are twinned, grow as aggregates, or are otherwise disordered. This has in some instances been overcome by reducing the rate of crystal growth, as was done with human lysozyme,[77] or by poisoning the mother liquor with some organic solvent. Dioxane seems to have provided the greatest success. Chymotrypsin was freed of a serious twinning problem by the addition of about 1% dioxane,[87] and phosphoglycerate kinase[100] by the addition of 3% dioxane to the mother liquor. It is not clear why dioxane has this effect, nor has a systematic investigation been made to discover other compounds with this property. It may be significant that Bailey also grew improved crystals of several plant-seed proteins from mixtures of ethanol and NaCl.[3,4]

9. Occasionally it is observed that a protein fails to crystallize after repeated trials and then suddenly, under otherwise identical conditions, begins to do so readily. The likely cause of this phenomenon may be ascribed to the fact that the protein sample was frozen and thawed several times as it was used to make trials, and that crystal formation only occurred after the process had been repeated half a dozen times. It is sometimes useful to try this if the protein refuses to form growth centers, or forms very twinned crystals.

10. A point that has not been seriously addressed is the decay or loss of scattering ability over time when protein crystals are exposed to X-rays. In some cases this damage is a major impediment to structure determination. While some studies of this phenomena have been carried out, no definitive explanation of its source has yet been offered. Cryogenic approaches

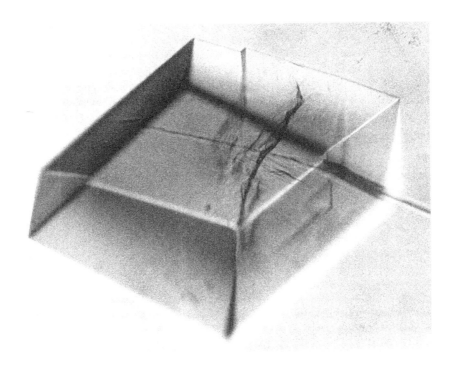

FIGURE 21.   A large flaw marks the upper face of this otherwise perfect rhombohedral canavalin crystal. The crack resulted from apparent mechanical stress after growth was complete, and was not originally present in the growing crystal. This was revealed by time lapse microphotography of growing specimens.

have to this time provided the best approach to the minimization of radiation damage[27,28] although other techniques involving the perfusion of the crystal with polymerizing agents have also been reported.[104] It appears that for crystals grown at low ionic strength or low concentrations of PEG, substantial increase of precipitant concentrations in the mother liquor may also afford some protection.[14]

As radiation sources, such as synchrotrons, having much more intense beams are employed, the problem of radiation damage may become more pronounced. This too is an area sorely in need of further experimentation and invention. In some cases the stabilization of crystals may prove a more unyielding problem than their growth.

11. Because the perfection, and perhaps also the terminal size, of a crystal depends to some extent on the orderly addition of molecules to growing faces, conditions at the crystal-liquid interface during growth may be of crucial importance. Chief among the influences at the interface are convective currents in the liquid produced by density gradients.[5] These are in turn a consequence of the depletion of protein from the adjacent mother liquor as molecules enlist in the solid state. These convective currents are gravity driven and may be quite severe in the case of rapidly growing crystals. In addition, crystal defects may occur in very fragile protein crystals simply from mechanical stress resulting from gravity as suggested by Figure 21. The question can then be asked: would crystals of greater size or perfection be grown in an environment where all molecular transport was purely diffusive, that is, in the absence of gravity where no convective currents exist.

To answer this question, experiments have been initiated by several groups of investigators to grow protein crystals in a gravity free, or "microgravity", environment. This means in space and, currently therefore, aboard the space shuttle once in orbit, or perhaps eventually aboard an orbiting space station. Several sets of experiments have already been

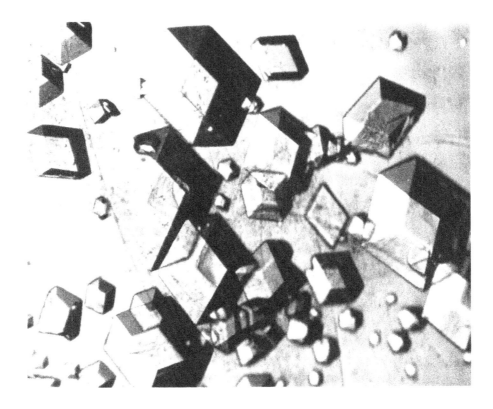

FIGURE 22. Rhombohedral crystals of canavalin, like those shown here, were grown in microgravity by vapor diffusion aboard the U.S. Space Shuttle from 0.7% NaCl at neutral pH.

conducted using a variety of proteins aboard the U.S. Space Shuttle. While the results of these experiments are still not definitive and much yet remains to be done, the initial results are highly encouraging.[20,54,55] Crystals of several proteins, one of which is shown in Figure 22, have been grown to large size in gravity free space, successfully returned to earth, and examined by X-ray diffraction analysis. These crystals have been shown to be definitely of as high a quality as those grown in laboratories on earth and indications are that they may have improved properties. This is still very preliminary, however, and additional experiments are certainly necessary.

REFERENCES

1. **Alderton, G. and Fevold, H. L.,** *J. Biol. Chem.,* 164, 1, 1946.
2. **Arnone, A., Bier, C. J., Cotton, F. A., Day, V. W., Hazen, E. E., Richardson, J. S., Richardson, D. C., and Yonath, A.,** *J. Biol. Chem.,* 24, 3202, 1971.
3. **Bailey, K.,** *Trans. Faraday Soc.,* 38, 186, 1942.
4. **Bailey, K.,** *Nature,* 145, 934, 1940.
5. **Baird, J. K., Meehan, E. J., Xidis, A. L., and Howard, S. B.,** *J. Cryst. Growth,* 76/3, 694, 1986.
6. **Baker, E. N. and Dodson, G.,** *J. Mol. Biol.,* 54, 605, 1970.
7. **Baldwin, E. T., Crumley, K. V., and Carter, C. W.,** *Proc. Biophys. Disc.,* November 10-13, 1985, Airlie, VA, Rockefeller Press, New York, p. 47.

8. Baranowski, T., *Z. Physiol. Chem.*, 260, 43, 1939.
9. Berthou, J., Leurent, A. L., Lebrun, E., and van Rapenbusch, R., *J. Mol. Biol.*, 882, 111, 1973.
10. Bott, R. R., Navia, M. A., and Smith, J. L., *J. Biol. Chem.*, 257, 9883, 1982.
11. Brayer, G. D. and McPherson, A., *J. Biol. Chem.*, 257, 3359, 1981.
12. Buckley, H. E., *Crystal Growth*, John Wiley & Sons, London, chap. 1, 1951.
13. Camerman, N., Hofmann, T., Jones, S., and Nyburg, S. C., *J. Mol. Biol.*, 44, 569, 1969.
14. Cascio, D., Williams, R., and McPherson, A., *J. Appl. Cryst.*, 17, 209, 1984.
15. Cohen, C. and Tooney, N. M., *Nature*, 251, 659, 1974.
16. Cohn, E. J., Hughes, W. L., and Weare, J. H., *J. Am. Chem. Soc.*, 69, 1753, 1947.
17. Cohn, E. J. and Ferry, J. D., *In Proteins, Amino Acids and Peptides*, Cohn, E. J., and Edsall, J. T., Eds., Reinhold, New York, 1950.
18. Creighton, T. E., *Proteins: Structures and Molecular Properties*, Freeman Co., New York, 1984.
19. Czok, R. and Bücher, Th., *Adv. Protein Chem.*, 15, 315, 1960.
20. DeLucas, L. J., Suddath, F. L., Snyder, R., Naumann, R., Broom, B., Pusey, M., Yost, V., Herren, B., Carter, D., Meehan, E. J., McPherson, A., Clifford, D., Walker, C., Nelson, B., Musgrave, S., Acton, L., and Bugg, C. E., *J. Cryst. Growth*, 76/3, 681, 1986.
21. Fenna, R. E., Matthews, B. W., Olson, J. M., and Shaw, E. K., *J. Mol. Biol.*, 84, 231-240.
22. Fitzgerald, P. M. D. and Madsen, N. B., *J. Crystal Growth*, 76/3, 600, 1987.
23. Fullerton, W. W., Potter, R., and Low, B. W., *Proc. Natl. Acad. Sci. USA*, 66, 1213, 1970.
24. Giege, R., Dock, A. C., Kern, D., Lorber, B., Thierry, J. C., and Moras, D., *J. Cryst. Growth*, 76/3, 554, 1986.
25. Granick, S., *J. Biol. Chem.*, 146, 451, 1941.
26. Green, D. W. and Aschaffenburg, R., *J. Mol. Biol.*, 1, 54, 1959.
27. Haas, D. J., *Acta Cryst.*, B24, 604, 1968.
28. Haas, D. J. and Rossmann, M. G., *Acta Cryst.*, B26, 998, 1970.
29. Hackert, M. L., Ford, G. C., and Rossmann, M. G., *J. Mol. Biol.*, 78, 665, 1973.
30. Hampel, A., Labanauskas, M., Conners, P. G., Kirkegard, L., RajBhandary, U. L., Sigler, P. B., and Bock, R. M., *Science*, 162, 1384, 1968.
31. Hanson, A. W., Applebury, M. L., Colman, J. E., and Wyckoff, H., *J. Biol. Chem.*, 245, 4975, 1970.
32. Hermans, J., *J. Chem. Phys.*, 77, 2193, 1982.
33. Hofmeister, T., *Arch. Exptl. Pathol. Pharmakol. Naunyn-Schmiedeberg's*, 24, 274, 1887.
34. Hopkins, F. G. and Pinkus, S. N., *J. Physiol.*, 23, 130, 1898.
35. Jacoby, W. B., *Anal. Biochem.*, 26, 295, 1968.
36. Jacoby, W. B., in *Methods in Enzymology, Vol. 11*, Jacoby, W. B., Ed., Academic Press, New York, 1971.
37. Jeng, T. W. and Chiu, W., *Ultramicroscopy*, 13, 27, 1984.
38. Jurnak, F., *J. Mol. Biol.*, 185, 215, 1985.
39. Jurnak, F. A. and McPherson, A., Eds., *Biological Macromolecules and Assemblies, Vol. II, Nucleic Acids and Interactive Proteins*, John Wiley & Sons, New York, Chap. 1-3, 1985.
40. Jurnak, F. A. and McPherson, A., Eds., *Biological Macromolecules and Assemblies, Vol. III, Active Sites of Enzymes*, John Wiley & Sons, New York, 1987.
41. Jurnak, F. A., McPherson, A., Wang, A. H. J., and Rich, A., *J. Biol. Chem.*, 255, 6751, 1980.
42. Kam, Z., Shore, H. B., and Feher, G., *J. Mol. Biol.*, 123, 539, 1978.
43. Kim, S. H., Quigley, G. J., Suddath, F. L., McPherson, A., Sneden, D., Kim, J. J., Weinzierl, J., and Rich, A., *J. Mol. Biol.*, 75, 421, 1973.
44. King, M. V., *J. Mol. Biol.*, 11, 549, 1965.
45. King, M. V., Bello, J., Pagnatano, E. H., and Harker, D., *Acta Cryst.*, 15, 144, 1962.
46. King, M. V., Magdoff, B. S., Adelman, M. B., and Harker, D., *Acta Cryst.*, 9, 460, 1956.
47. Knox, J. R., Zorsky, P. E., and Murthy, N. S., *J. Mol. Biol.*, 79, 597, 1973.
48. Kraut, J., Matthews, D. A., Alden, R. A., Bolin, J. T., Freer, S. T., Hamlin, R., Xuong, N., Poe, M., Williams, M., and Hoogsteen, K., *Science*, 197, 452, 1977.
49. Kunitz, M., *J. Gen. Physol.*, 35, 423, 1952.
50. Kuo, I. A. M. and Glaeser, R. M., *Ultramicroscopy*, 1, 53, 1975.
51. Lagerkvist, U., Rymo, L., Lindquist, O., and Anderson, E., *J. Biol. Chem.*, 247, 3897, 1972.
52. Lee, J. C. and Lee, L. L. Y., *J. Biol. Chem.*, 256, 625, 1981.
53. Liljas, A. and Rossmann, M. G., *Ann. Rev. Biochem.*, 43, 475, 1974.
54. Littke, W. and John, C., *Science*, 225, 203, 1984.
55. Littke, W. and John, C., *J. Cryst. Growth*, 76/3, 663, 1986.
56. Low, B. W., Chen, C. C., Berger, J. E., Singman, L., and Pletcher, J. F., *Proc. Nat. Acad. Sci. (U.S.A.)*, 56, 1746, 1966.
57. Matthews, B. W., *J. Mol. Biol.*, 33, 491, 1968.

58. **McMeekin, S. L.**, *J. Am. Chem. Soc.*, 61, 2884, 1939.
59. **McPherson, A.**, *J. Biol. Chem.*, 251, 6300, 1976.
60. **McPherson, A.**, *Methods of Biochemical Analysis, Vol. 23*, Glick, D., Ed., Academic Press, New York, 1976, 249.
61. **McPherson, A.**, *The Preparation and Analysis of Protein Crystals*, John Wiley & Sons, New York, 1982.
62. **McPherson, A.**, The crystallization of macromolecules: General principles,, in, *Methods in Enzymology: Diffraction Methods*, Hirs, M., Timasheff S. N., and Wyckoff, H., Eds., Academic Press, New York, 114, 112, 1985.
63. **McPherson, A., Friedman, M. L., and Halsall, H. B.**, *Biochem. Biophys. Res. Commun.*, 124, 629, 1984.
64. **McPherson, A., Koszelak, S., Axelrod, H., Day, J., Willisma, R., Robinson, L., McGrath, N., and Cascio, D.**, *J. Biol. Chem.*, 261, 1969, 1986.
65. **McPherson, A. and Rich, A.**, *Biochem. Biophys. Acta*, 285, 493, 1972.
66. **McPherson, A. and Rich, A.**, *J. Ultrastruct. Res.*, 44, 75, 1973.
67. **McPherson, A. and Shlichta, P.**, *J. Cryst. Growth*, 1987.
68. **McPherson, A. and Spencer, R.**, *Arch. Biochem. Biophys.*, 169, 650, 1975.
68a. **McPherson, A., Mickelson, K. E., and Westphal, U.**, *J. Steroid Biochem.*, 13, 991, 1980.
69. **Miers, H. A. and Isaac, F.**, *J. Chem. Soc.*, 89, 413, 1906.
70. **Miers, H. A. and Isaac, F.**, *Proc. Roy. Soc.*, A79, 322, 1907.
71. **Moews, P. C. and Bunn, C. W.**, *J. Mol. Biol.*, 54, 395, 1970.
72. **Moews, P. C. and Bunn, C. W.**, *J. Mol. Biol.*, 68, 389, 1972.
73. **Northrop, J. H.**, *Ergeb. Enzymforsch.*, 1, 302, 1932.
74. **Northrop, J. H., Kunitz, M., and Herriott, R. H.**, *Crystalline Enzymes*, Columbia University Press, New York, 1948.
75. **Osborne, T. B.**, *Am. Chem. J.*, 14, 662, 1892.
76. **Osborne, T. B.**, *The Vegetable Proteins*, 2nd ed., Longmans, Green, London, 1924.
77. **Osserman, E. F., Cole, S. J., Swan, I. D. A., and Blake, C. C. F.**, *J. Mol. Biol.*, 46, 211, 1969.
78. **Petrov, T-G., Treivus, E. B., and Kasatkin, A. P.**, *Growing Crystals from Solution*, trans. from the Russian, Consultants Bureau, Plenum Press, New York, 1969.
79. **Pusey, M. L.**, *Anal. Biochem.*, 158, 50, 1986.
80. **Pusey, M. L. and Nauman, R.**, *J. Cryst. Growth*, 76/3, 593, 1986.
81. **Quiocho, F. A. and Richards, F. M.**, *Biochemistry*, 5, 4062, 1966.
82. **Ray, W. J. and Bracker, C. E.**, *J. Cryst. Growth*, 76/3, 562, 1986.
82a. **Ray, W. J. and Prevathingal, J. M.**, *Anal. Biochem.*, 1986.
83. **Rayment, I.**, *Meth. Enz. Vol.*, 114, 136, 1985.
84. **Rosenberger, F.**, *J. Cryst. Growth*, 76/3, 618, 1986.
85. **Salemme, F. R.**, *Arch. Biochem. Biophys.*, 151, 533, 1972.
86. **Schulz, G. E. and Schirmer, R. H.**, *Principles of Protein Structure*, Springer-Verlag, New York, 1979.
87. **Sigler, P. B., Jeffery, B. A., Matthews, B. W., and Blow, D. M.**, *J. Mol. Biol.*, 15, 175, 1969.
88. **Solomon, A., McLaughlin, C. L., Wei, C. H., and Einstein, J. R.**, *J. Biol. Chem.*, 245, 5289, 1970.
89. **Sorensen, S. P. L. and Hoyrup, M.**, Studies on proteins. C.-R. Lab. Carlesberg 12, 12, 1915 to 1917.
90. **Steitz, T. A., Fletterick, R. J., and Hwang, K. J.**, *J. Mol. Biol.*, 78, 551, 1973.
91. **Sumner, J. B. and Somers, G. F.**, *The Enzymes*, Academic Press, New York, 1943.
92. **Sumner, J. B. and Somers, G. F.**, *Laboratory Experiments in Biological Chemistry*, Academic Press, New York, 1944.
93. **Thaller, C., Eicher, G., Weaver, L. H., Wilson, E., Karlsson, R., and Jansonius, J. N.**, *Meth. Enz. Diffraction Methods for Biological Macromolecules*, pp. 132, 1985.
94. **Thaller, C., Weaver, L. H., Eichele, G., Wilson, E., Karlsson, R., and Jansonius, J. N.**, *J. Mol. Biol.*, 147, 465.
95. **Tsutomu, A. and Timasheff, S. N.**, *Meth. Enz., Diffraction Methods for Biological Macromolecules*, 114, 49, 1985.
96. **Unwin, P. N. T. and Henderson, R.**, *J. Mol. Biol.*, 94, 425, 1975.
97. **Vainshtein, B. K.**, *Growth of Crystals*, Vol. 13, Givargizov, E. I., Ed., Acad. Sciences of USSR, Moscow. Trans. from Russian, Nauka Press, 1986.
98. **Vickery, H. B., Smith, E. L., Hubbell, R. B., and Nolan, L. S.**, *J. Biol. Chem.*, 140, 613, 1941.
99. **Waller, J. P., Risler, J. L., Monteilhet, C., and Zelwer, C.**, *FEBS Lett.*, 16, 186, 1971.
100. **Watson, H. C., Wendell, P. L., and Scopes, R. K.**, *J. Mol. Biol.*, 57, 623, 1971.
101. **Weber, B. A. and Goodkin, P. E.**, *Arch. Biochem. Biophys.*, 141, 489, 1970.
102. **Wyckoff, H. W., Tsernoglou, D., Hanson, A. W., Knox, J. R., Lee, B., and Richards, F. M.**, *J. Biol. Chem.*, 245, 305, 1970.
103. **Yang, H. J., Lee, B., and Haslam, J. L.**, *J. Mol. Biol.*, 81, 517, 1973.

104. **Zaloga, G. and Sarma, R.,** *Nature,* 251, 551, 1974.
105. **Zeppenzauer, M., Eklund, H., and Zeppenzauer, E.,** *Arch. Biochem. Biophys.,* 126, 564, 1968.
106. **Zeppenzauer, M.,** Formation of large crystals, in *Methods of Enzymology,* Jacoby, W. B., Ed., Academic Press, New York, 1971 22, 253.

Chapter 2

DETERGENT PHENOMENA IN MEMBRANE PROTEIN CRYSTALLIZATION

Martin Zulauf

TABLE OF CONTENTS

I. INTRODUCTION

Membrane proteins are amphiphilic macromolecules incorporated vectorially in the lipid membrane in a quasi-solid state. Part of their surface that is in contact with the lipid bilayer core is hydrophobic, while other parts that are exposed to the aqueous environment on either side of the membrane are hydrophilic. Upon disruption of the membrane and removal of the lipids, such proteins tend to aggregate unspecifically and to precipitate in bulk. This can be avoided, and the proteins can be solubilized in aqueous solutions, when the membrane lipids are replaced by detergents[1,2] which, at the concentrations used in biochemistry, do not form membrane-like structures.

Solubilized membrane proteins form so-called mixed micelles with the detergent. Under appropriate conditions, their hydrophobic surface parts are buried in the interior of a micelle or are covered by a (curved) monolayer film of detergent. Depending on the size of the protein and the organization of the exposed amino acids, mixed micelles may therefore have many different geometries, ranging from almost unperturbed micelles (diameter 40 to 60 Å), into which the protein is embedded and from where it may or may not protrude, to more complex objects with surfaces made up of protein and detergent coats (thickness 20 to 30 Å). The designation "mixed micelle" may not always be appropriate for what a molecular observer would see in solution: instead of looking like essentially normal micelles, mixed micelles may sometimes appear as abnormally large micelles of unusual geometry or even as normal protein with a minor micelle attached to it.

Complete solubilization of a membrane protein is achieved when the protein is not further complexed by the detergent, i.e., when there is not more than one protein in a mixed micelle. To avoid such a situation of incomplete and evasive solubilization, high enough detergent concentration should be chosen such that the free-to-mixed micelle ratio is at least ten.[3] This is usually achieved when detergent concentration is around a few times the critical micelle concentration (CMC); see below. A pictorial description of a completely solubilizing solution is then a solution which contains a lot of normal (empty) micelles, detergent monomers and mixed micelles, the latter all of the same shape, depending on the protein.

When such solutions are subject to crystallization assays devised in the same way as with soluble proteins, new phenomena may occur which are particular to the detergents. In the same way as water soluble proteins can be salted out (possibly into crystals) by precipitating agents, also micelles can be salted out. They do not form crystals but usually phase separate: the single phase detergent solution becomes biphasic, the two phases being macroscopically discernible, and one of them contains most of the micelles (also mixed micelles, if present). This process can be studied independently of the solubilized proteins. Phase separations of this kind occur naturally in particular detergent systems even in the absence of precipitating agents, and they have therefore been examined by workers in the colloid field.

Crystals of membrane proteins contain, as a constituent part, the detergents bound to the protein and necessary for the solubilization; when one tries to dialyze the detergent out of the crystals, they dissolve or collapse (even after fixation with glutaraldehyde or so). In the crystals the protein occupies crystallographically defined positions, the detergent does not. In X-ray crystallography, the detergent may not even produce an easily recognizable density in the maps (see results on the reaction center), and one has to recur to neutron diffraction in particular contrast situations to visualize the detergent in the crystals. Nevertheless, the crystals may contain detergent at substantial concentrations (compared to crystal water), far higher than a few times the CMC (as necessary for complete solubilization) and possibly high enough that mesophases may occur. Under certain circumstances mesophases might interfere with crystal growth. It seems to be useful if the biochemist acquires some knowledge of mesophases formed by detergents, and therefore they will be discussed below.

Much of the concepts invovled in membrane protein solubilization (lipid removal, detergent binding to proteins, measuring protein properties in the presence of detergents, etc.) have been studied and described in excellent reviews.[1,2,9] They will not be repeated here. Also, excellent articles on colloidal properties of detergents exist.[4-6] The book by Attwood and Florence[7] is a rich source of information of surfactants, both from the physical chemical point of view as well as with respect to applications in biology and pharmacy (it contains over 2000 references).

It is impossible to present a satisfactory review of such a field in a few pages. I have concentrated on a few topics in detergent phenomenology which seem to me relevant to understand phenomena encountered in membrane protein crystallization and which to my knowledge have not been presented in this context before.

After a brief survey of fundamentals, I will discuss some aspects of mesophases and naturally occurring phase separations in polyoxyethylene detergents. An effort is made to suggest molecular mechanisms that underly these phenomena. The basic concepts are that micelles are strongly interacting objects. The nature of the interactions is discussed, although it may not be established unambiguously at the present time. I hope that the concepts may be useful to the reader who tries to crystallize membrane proteins, by providing a rationale for the interpretation of the phenomena which he or she observes.

II. CLASSIFICATION OF SURFACTANTS

Surface active agents (surfactants) are molecules formed from a polar entity (the hydrophilic head group) and nonpolar aliphatic or aromatic compounds (the hydrophobic tail). Being amphiphilic, some of the surfactant molecules will always dissolve in water as monomers, although the bulk material will remain complexed in some way. At the interface of water with air (or another liquid, an oil), the monomers then tend to orient themselves such that the polar head remains in the water but the tail protrudes into the air, thereby changing surface properties such as the surface tension of water (hence their name). This ability distinguishes surfactants from strictly nonsoluble materials such as, e.g., octadecane, cholestane, carotene, etc. The solubility of the bulk material, however, is different for different surfactants which allows for their classification.[5,8] When water is added to solid surfactants, e.g., by putting small amounts of dry material between glass coverslips under a polarizing microscope, to which a drop of water is added at the edge of the coverslip, three types of behavior can occur:

1. The surfactant is practically insoluble, the solution contains only solids, some monomers, and water (nonswelling amphiphiles). This is observed for polar and semipolar lipids, in particular for the majority of lipid substances in animals, di- and triglycerides, long chain protonated fatty acids and esters thereof, long chain alcohols, cholesterol, retinols, vitamins A, K, and E, etc.
2. Although the substances remain virtually insoluble except for the monomers, they swell and lyotropic liquid crystals form, but no micelles. Lyotropic crystals are easily identified under the microscope as birefringent (bright and colored in the dark field between crossed polarizers). This is found for the lipids of the cell membranes (phospholipids, glycolipids, cerebrolipids, sphingomyelin) and some monoglycerides.
3. The material dissolves completely to form isotropic, homogeneous solutions containing micelles and substantial amounts of monomers. Liquid crystals may occur at higher concentrations. Examples are anionic, cationic, nonionic, and zwitterionic detergents, gangliosides, bile salts, saponins, lysolecithin, etc.

Solubility properties depend crucially on temperature and concentration. Thus, some detergents behave below a well-defined temperature (the so-called Krafft point) as if they were nonswelling amphiphiles, but as normal detergents above this tempereature. Phospholipids are swelling amphiphiles only at temperatures high enough to render the aliphatic chains partly liquid, which depends on chain length, saturation, branching, and substitution. Therefore chemically related substances may belong to different categories, depending on conditions. The distinction of the categories should not be understood in a strict sense, and particular compounds which belong chemically to a class that is identified as, say, nonsoluble swelling amphiphiles, may nevertheless occasionally be soluble and detergent-like.

Of all these surfactants we will concentrate, in the following, on nonionic and zwitterionic detergents with an alkyl tail, because they are the detergents most often used for membrane protein solubilization and crystallization. They are called ''mild'' detergents, because they usually do not denature proteins. Historically, detergents such as Triton X-100 were provided to biochemists by the laundry industry. Nowadays, homologous series of them are commercially available. A few of them are also available with the alkyl exchanged for a steroid (e.g., the sulfobetaines, commercialized under the name of CHAPS® and CHAPSO®).

A. MICELLES AND THE CRITICAL MICELLE CONCENTRATION

At low concentrations, detergents in water form small colloidal aggregates called micelles. The apolar tails of the amphiphilic molecules are stowed away in the interior of the micelle structure, and the polar heads face the water.

The effects which drive detergents into micellization are related to the structure of water, and thus micelle formation is a manifestation of the hydrophobic effect.[9] According to Nemethy and Sheraga[10] structured regions continuously form and break down in bulk water, where the water molecules are hydrogen-bonded together in a tetrahedral network similar to that of ice. A detergent molecule locally distorts that network. In particular, its hydrophobic tail cannot participate in an already existing cluster. Around the tail, water has to restructure in an appropriate arrangement which must be maintained permanently (''hydrophobic hydration''). This decreases entropy, making dissolution of the hydrocarbon part unfavorable. At the same time, internal torsional vibrations of the hydrocarbon are restricted because of the water cage. Removal of the hydrophobic tail of a detergent molecule from water is entropically advantageous, because structured water is released and mobility constraints are abolished. Removal is possible either by insertion of the molecule in the air/water or air/oil interface, or by self-aggregation into colloidal structures; the micelles.

Micelle diameters are approximately 5 nm (exceptioanlly rod or disc-like micelles exist), forming solutions which are transparent, isotropic, and of low viscosity. However, micelles can only exist above a particular detergent concentration (10^{-1} to 10^{-6} M) which is the critical micelle concentration (CMC). Above the CMC, micelles always coexist with unaggregated monomers, present at a concentration approximatley equal to the CMC. Thus, the concentration of micelles and monomers varies with total detergent concentration as shown in Figure 1. Micelles are colloidal particles, i.e., they are by nature highly dynamic, with continuous exchange of monomers between them and the nonaggregated molecules.

The full lines in Figure 1 illustrate a somewhat idealized situation. Two factors may become critical. First, the monomer concentration may not remain constant above the CMC, but rather continue to drop moderately (not as little as shown in the figure). This is the case for some detergents with steroid tails (taurocholate,[48] saponins). Second, the onset of micelle formation is not abrupt, but occurs gradually over an extended concentration interval (broken curve in Figure 1). The interval increases with decreasing detergent size. The two factors lead to an inherent uncertainty in the determination of the CMC, which is rarely defined better than up to 5 to 10%. Incidentally, published data on CMCs of common detergents scatter about this amount.

The value of the CMC depends on the specific detergent considered. Although a good

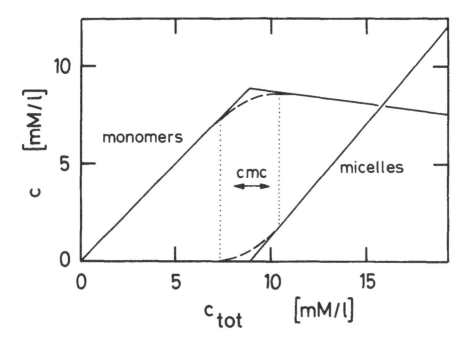

FIGURE 1. Monomer and micelle concentrations as a function of total detergent concentration, c_{tot}, for sodium dodecylsulfate in water at room temperature. (Adapted from Gunnarsson, G., Jönsson, B., and Wennerström, H., *J. Phys. Chem.*, 84, 3114, 1980.)

''balance'' between the size of the alkyl tail and the hydrophilic head group is an important factor for the solubility (see below), it is the length of the alkyl chain which effectively determines the value of the CMC. A rule of thumb derived from Traube's rule on surface tension maintains that if a detergent with m carbon atoms in the alkyl chain has a CMC of a given value, the CMC of a homologue with $m - 1$ carbons is roughly three times this value, for $m - 2$ carbons roughly ten times.

Table 1 shows CMCs of a series of homologues of polyoxyethylene monoethers; note that the rule breaks down for very short and very long alkyl tails. In the spirit of the rule, a phenyl is worth 3.5 methylenes (Triton series, Table in appendix of the book). Sometimes the rule does not apply to detergents of industrial use (e.g., Tween), because these products are often molecularly heterogeneous. Experimental ways to determine CMCs will be discussed below.

Although the concept of micelles is old and the theory of micelle formation has been developed to a considerable degree of sophistication (reviewed in Reference 4 and 7), recent years have seen a renewed interest in them. The interest has concentrated on the topics listed below.

1. Micellar Size and Shape

Although most micelles can be considered as ''essentially spherical'', it has been pointed out that this cannot be strictly true. The micellar radius should not exceed that length of the maximum possible extension of the detergent molecule. Yet, experimentally determined radii are usually 10 to 30% larger. Tanford[11] has proposed that the micellar shape should be considered as an ellipsoid of revolution, with the minor semi-axes being constrained to the monomer length. Axial ratios are usually small (between 1 and 2), and the strong curvature of the ''end'' or ''rim'' inherent in this shape has been criticized.

2. Micellar Hydration

Measurements with nmr and esr have revealed that most of the methylene groups of the

TABLE 1
Critical Micelle Concentrations (CMC)
and Cloud Point Temperatures (T_d) for
Alkyl Polyoxyethylene Monoethers

C_mE_n		CMC in H_2O		
m	n	(mM)	(% w/w)	$T_d(°C)$
16	21	0.0039	0.00032	>100
	12	0.0023	0.00018	92
	8	0.0021	0.00012	63
14	8	0.009	0.00051	—
12	8	0.071	0.0038	77
	6	0.063	0.0028	48
10	9	1.3	0.072	—
	6	0.9	0.038	59
8	5	6.0	0.21	63
	4	7.2	0.22	42
	3	—	—	10
7	5	21.0	0.7	73
	4	—	—	51
6	5	37.0	1.2	—
	3	23.0	1.1	51

Note: The CMC's were determined by the drop volume method. T_d for m ≥ 10 from reference 6, for m < 10 as measured for 1% solutions (2% for m = 6).

alkyl chain may sometimes be in contact with water. This was interpreted as being due to water penetration into the core. Recent theoretical and experimental investigations into the structure of the core have stressed the flexibility of the alkyl chains and shown that even the terminal methylene group has a chance to be positioned at the core surface, thereby entering in contact with water.[12,13]

3. Micellar Polydispersity

Most workers in the field know that micellar systems are "quite monodisperse", but exhibit some polydispersity. The quantitative analysis is difficult. It consists of measuring the weight average and number average molecular weight of the micelles. The latter has eluded direct measurement, as vapor pressure osmometry is not sensitive enough. Polydispersity analysis of quasielastic light scattering data is particularly difficult for unimodal distributions, and observations of the asymptotic scattering in small angle techniques requires substantial surfactant concentrations, posing additioinal problems.

4. Intermicellar Interactions

Only recently has it been realized that both ionic and nonionic micelles may have strong interactions, repulsive for the former (due to Coulomb repulsion) and attractive for the latter. This point will be discussed in detail below.

5. Micellar Kinetics

Micelles are dynamic objects which form and decay on a slow time scale (millisecond range), and which exchange monomers with the surrounding bulk solution on a fast time scale (< 10 μs depending on the CMC). The latter process has been detected by ultrasonic absorption[14,15] and has been modeled theoretically, but its influence on micelle structure has not been established conclusively. In many structural analyses, micelles are generally treated as if they were rigid macromolecules.

B. MESOPHASES

Depending on concentration and temperature, detergents may not only form micelles in water, but also aggregates of extended, spatially ordered structure termed mesophases. Table 2 lists the phases most commonly encountered, and gives typical dimensions. The most important of the ordered phases, the hexagonal and lamellar phases (H_1 and L_α in the notation of Tiddy[5]) can easily be identified in the lab. They do not flow in glass tubes (high viscosity), and they are birefringeant (they appear bright and colored in the dark field between crossed polarizers). They can also be distinguished by the texture seen under the microscope;[17] however, unambiguous assignment based on optical examination remains somehow an art and demands long experience. To identify them properly, one has to recur to small angle X-ray diffraction and nmr spectroscopy.

A convenient way to discuss mesophases of detergents is to look at so-called phase diagrams, which are temperature vs. concentration diagrams into which boundary lines are drawn delineating the phases. Figure 2 shows some phase diagrams of polyoxyethylene monoethers (Table 1). These detergents are chosen because they are the best studied of all mild detergents.[6] Actually, the diagrams in Figure 2 are simplified. Phase boundaries, particularly towards the mesophases, often have a finite extension defining zones where adjacent phases coexist. An example of the complexity is discussed in Reference 18.

It is profitable to examine Figure 2 with some care. Consider, e.g., $C_{12}E_6$. At low temperature, the micellar phase exists up to roughly 35% w/w. Denser packing of the micelles by removing water is impossible. If one tries, micelles escape closer packing by rearranging into very long cylinders, and the amphiphile forms a hexagonal phase H_1 which can obviously cope with less water between the aggregates than micelles can. Upon further removal of water, the surfactant rearranges a second time in a L_α, with the water intermediate between extended detergent bilayers. The transition is accompanied with an intermediate phase, a bicontinuous V_1. Although the L_α could conceptually cope with all water being removed between the bilayers (such that they would touch each other), this does not happen. Above 80% detergent or so, the ordered structure is lost and a structureless liquid L of decreasingly hydrated amphiphile is found.

Mesophases may be distorted, particularly lamellar phases. Mono- and multilamellar vesicles (liposomes) are examples. Thus, they sometimes appear less clear, but rather hazy. Note also that the mesophases are heat sensitive; at a well-defined temperature the phases melt (samples start to flow and lose birefringeance). As seen in the diagram for $C_{12}E_6$, there is a path where the micellar phase is not bounded when concentration is increased. The homogeneous, transparent L_1 of low viscosity persists up to the pure surfactant liquid L, without an apparent boundary between L_1 and L. Solution viscosities go through a maximum at around 75%, but always remain comparable to that of water (a factor of 10 or higher). What actually happens is that the number of micelles decreases gradually, and the number of (hydrated) monomers increases correspondingly, so the distinction between L_1 and L is somewhat artificial at elevated concentrations.

There is another line shown in this diagram, the one separating L_1 from $L_1' + L_1''$. It is connected with a phase separation, or consolution phenomenon, which we discuss in Section II.D.

Consider now the diagram of C_8E_5 and $C_{16}E_8$. They are akin to that of $C_{12}E_6$, except that the gap between the consolute boundary and the mesophases is larger for the former and smaller for the latter. For $C_{16}E_8$, the L_α has become more prominent, and at low temperature a cubic I_1 develops (see Table 2). From this we learn that for detergents which form micelles at low concentration, the sequence of mesophases encountered with increasing concentration is

$$L_1 \rightarrow I_1 \rightarrow H_1 \rightarrow V_1 \rightarrow L_\alpha \rightarrow L/S$$

with omissions depending on temperature and surfactant.

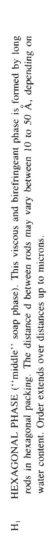
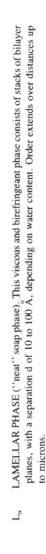

TABLE 2
Structure of Detergent Phases in Water

M   MONOMERS are ubiquitous.

L₁   MICELLES may have spherical, rod or disc shape. The minimum dimension exceeds twice the all-trans length of the molecules by 10 to 30%.

H₁   HEXAGONAL PHASE ("middle" soap phase). This viscous and birefringeant phase is formed by long rods in hexagonal packing. The distance d between rods may vary between 10 to 50 $\mathring{A}$, depending on water content. Order extends over distances up to microns.

L$_\alpha$ LAMELLAR PHASE ("neat" soap phase). This viscous and birefringeant phase consists of stacks of bilayer planes, with a separation d of 10 to 100 $\mathring{A}$, depending on water content. Order extends over distances up to microns.

I   CUBIC PHASE (viscous isotropic). This very viscous phase is non-birefringeant and is formed by spherical micelles in a cubic crystalline lattice.

L$_\beta$ GEL PHASE. This phase is viscous and non-birefringeant. Hydrocarbon chains form a crystalline arrangement.

V₁   BICONTINUOUS PHASE. Water and detergent channels interpenetrate; any two points in one of them may be connected by a path situated entirely within that moiety.

S,L   PURE, HYDRATED DETERGENT. This may be a structureless liquid or a solid (waxy, crystalline).

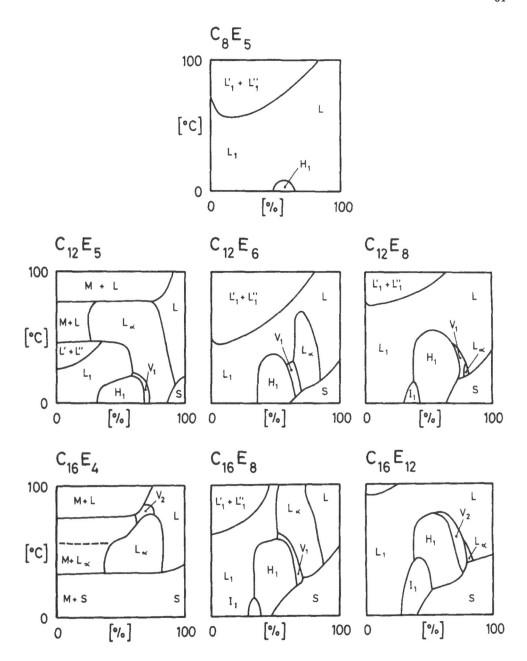

FIGURE 2. Phase diagrams of seven polyoxyethylene monoether detergents in H₂O. The boundaries in these plots showing temperature T vs. concentration w (% w/w) distinguish the various phases. Notation of phases according to Table 3. (Partially adapted from Mitchell, D. J., Tiddy, G. J. T., and Waring, L., et al., *J. Chem., Soc., Faraday Trans. 1*, 79, 975, 1981)

The next diagrams to consider might be those for $C_{12}E_8$ or $C_{16}E_{12}$. Note that L_α almost disappears, but I_1 becomes more prominent. Also, the consolute boundary has moved to higher temperatures. On the other hand the phase diagrams for $C_{12}E_5$ and $C_{16}E_4$ differ dramatically. The micellar phase L_1 has shifted to lower temperatures or disappeared. Obviously the head of these surfactants does not balance the tail and according to the classification in Section II these detergents are swelling amphiphiles. This also manifests from the appearance of several regions where 2 phases coexist: $L_1' + L_1''$, $L_1' + L_\alpha$, or $L_1' +$ L (L_1' stands for a micellar phase with concentration very near to the CMC).

C. SURFACE TENSION AND THE DETERMINATION OF THE CRITICAL MICELLE CONCENTRATION

As mentioned in Section II, the ubiquitous monomers in surfactant solutions tend to orient themselves at the air/water interface. By their intrusion into the interfacial water layer, the surface tension is reduced. Intermolecular forces between water and detergent molecules are generally less than those existing between water molecules, therefore the contracting power of the surface is reduced.

For detergents, the surface film is dynamic: monomers enter and leave the interface in proportion to the monomer concentration in solution. (This is not the case for the swelling and nonswelling surfactants, which form stable films which develop only gradually in time.) Below the CMC, monomers constitute the only solute material, and surface tension varies in relation to their concentration. When the CMC is reached, additional surfactant put in solution forms micelles and the monomer concentration remains more or less constant. Due to micellization, a crowding of the interface is avoided and the reduction of surface tension is limited. We therefore expect the surface tension to decrease gradually, as a function of total surfactant concentration, from its value for pure water (72 dyn/cm), and to attain a plateau at and above the CMC. This is actually observed in experiments, as shown in Figure 3. Typically, detergents lower the surface tension of water in air by $1/3$ to $1/2$. The free monomer concentration of lipids, being orders of magnitude smaller than the CMC of detergents (10^{-8} to 10^{-12} M), has correspondingly a minor effect on surface tension.

It follows that the CMC of a detergent can be determined from surface tension measurements of a series of solutions with ascending concentrations extending above the CMC. A value for the latter may be obtained qualitatively by shaking the samples vigorously so that foam forms. Foam remains more stable and for a longer time above than below the CMC. Another qualitative test consists in placing or spilling a drop of detergent solution on a (hydrophobic) parafilm sheet, and observing contact angles or how easily the droplets splash.[19]

There are many excellent methods to measure surface tension accurately.[20] As discussed above, the CMC is in practice not a sharply defined concept, so when one deduces it from surface tension measurements, a simple and even somewhat imprecise method will suffice. The simplest method is based on the observation that drops forming at the orifice of a tube cannot exceed a maximum size before they fall. Drops that hang at the orifice because their weight is balanced by forces acting along the edge where the drop attaches to the tube. That force per unit length of the edge is the surface tension. It follows that the maximum drop size is proportional to the surface tension. An easy way to perform drop volume measurements consists in equipping a syringe with a precisely machined, wettable outlet tube (e.g., brass). Successive drops may be driven out of the syringe by moving the piston, controlled, e.g., by turning it in a screw. Thus, the number of turns necessary for successive drops to detach at the orifice is monitored for a series of concentrations. This leads to a plot such as the one shown in Figure 3 (right hand scale). The CMC is then read off from the intersection of two straight lines, one in the descending part of the curve, the other through the plateau. The construction and operation of a little device to perform such measurements with a minimum of effort has been described elsewhere.[19]

It is useful to perform surface tension measurements when working with a new detergent whose properties are not well known. It allows one to assess impurities in the product. Trace amounts of solvents, e.g., may cause noticeable effects. Instead of a neat intersection of the descending and plateau curves, an undershoot at the CMC may occur before the plateau is reached. This is usually accompanied by abnormal micelle size near the CMC (as measured,

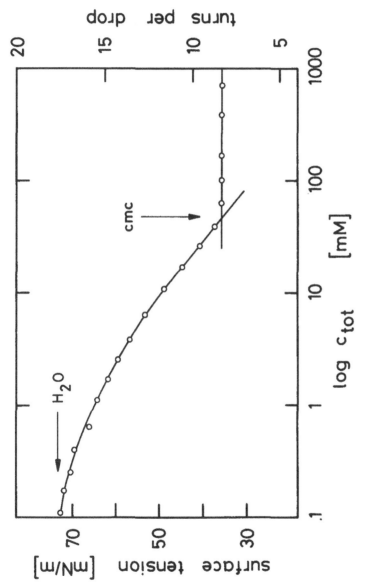

FIGURE 3. Surface tension measurements of dodecyl octaoxyethylene monoether ($C_{12}E_8$) in 150 mM NaCl at room temperature. Left hand ordinate: surface tension (1 mN/m = 1 mN/m = 1 dyn/cm = 1 g/s²). Right hand ordinate: number of turns necessary to move the piston such that a successive drop detaches at the orifice (c.f. text). Data are plotted as a function of the logarithm of the detergent molarity. The CMC is obtained from the intersection of the final slope with the plateau extrapolation.

say, by photon correlation spectroscopy), possibly due to solubilization of the solvents in droplets larger than the size of micelles. In these instances, the detergent ought to be purified. Also, commercial detergents of industrial use are often heterogeneous in alkyl chain length, which may lead to several shoulders in the surface tension curve.

A further benefit from careful surface tension measurements is that the surface area occupied by a single detergent molecule can be calculatead from the slope of the surface tension curve right before the intersection with the plateau.[7,21]

D. THE CONSOLUTION BOUNDARY OF POLYOXYETHYLENE DETERGENTS

We now return to the phase diagrams of Figure 2, and turn our attention to the consolution boundary, the line separating the micellar phase L_1 from $L_1' + L_1''$. The underlying phenomenon is as follows. When a micellar solution is slowly heated up, a temperature will eventually be hit where the sample suddenly turns turbid ("cloud point"). The clouding temperatures as a function of detergent concentration constitute the (lower) consolution boundary. When the solution is brought to a temperature above the cloud point and kept there for a while, the sample clears up and separates into two transparent, isotropic solutions discernible by a well visible meniscus. Both solutions are micellar (hence the designation $L_1' + L_1''$), but one contains almost all detergent, whereas the other is slightly above the CMC. This is visualized in Figure 4, which shows a construct by which the detergent concentration in the two phases can be calculated. The consolution boundary is thereby considered as a coexistence curve.

This phase separation is a peculiarity of all polyoxyethylene detergents (Table 1). You may easily convince yourself that, e.g., a 5% Triton X-100 solution clouds at 66°C. The cloud point may exceed 100°C for surfactants with long ethylene oxide chains. It can then be measured under pressure (sealed tubes) or with increasing amounts of additives that lower the critical point (see below), and subsequent extrapolation to zero additive.

The nature of this phase transition has been investigated in detail.[22-26] In the following, I present a simplified explanation why the phase separation occurs. It is based on the concept of intermicellar interactions, assuming that micelles do not change their small, globular shape during the process. This point has been challenged,[27-29] but I do not enter into a discussion of the controversy, which is partly related to experimental techniques and tends now to settle in favor of interactions rather than micellar growth.

Polyethylene oxide is a hygroscopic substance. In aqueous solutions, the polymers are highly hydrated. Each ether oxygen may bind several molecules of water, which it carries with it on Brownian motion. Polyoxyethylene micelles therefore contain a hydrocarbon core made up by the alkyl chains, surrounded by coiled oligomers of ethylene oxide and a lot of hydration water. When two micelles collide, the bound water acts like an ice shield which keeps the two cores well separated. These cores, made of material with a dielectric constant of $\epsilon = 2$, would well develop dispersion forces,[22] since they are separated by water with $\epsilon = 80$. However, these forces decay very rapidly with distance (van der Waals forces are short-range). Now when the temperature is raised, some of the hydration water "melts", allowing closer contact of the micelles upon collision and therefore stronger sampling of the dispersion forces. Thus, micelles become attractive at elevated temperatures. The attractive intermicellar potential has a range much smaller than the radius of the micelles: only upon close encounters do they have a tendency to "stick". The interaction energy is of the order of kT,[22,26] thus the temporary binding of two micelles is easily broken by thermal energy. However, this short-range potential is enough to induce long-range effects in the solution. Instead of the micelles being spatially distributed evenly in the solvent, they start to form loose, open clusters which are separated by regions containing micelles at low concentration. The attractive potential therefore leads inevitably to concentration fluctuations,

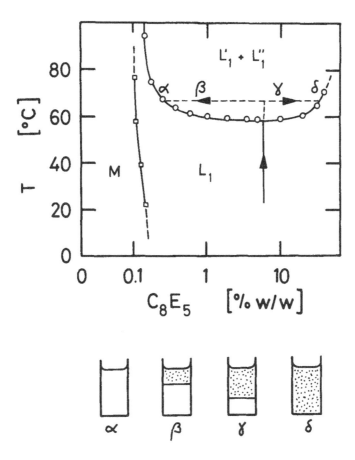

FIGURE 4. Phase diagram of C_8E_5, temperature T versus concentration, in semilogarithmic representation. Data indicate that the consolution boundary does not intersect with the CMC line (no phase separation occurs below the CMC). To estimate the concentration of the two phases obtained after heating a micellar solution above the cloud point, draw the broken horizontal line corresponding to the final temperature. It intersects the consolution boundary at two points with concentrations approximately equal to that of the two phases (the consolution boundary can be considered as a coexistence curve). The volume of the two phases can be estimated from the total initial detergent concentration. When this was lower than the critical concentration corresponding to the minimum cloud point temperature (points α and β), the concentrated phase is of smaller volume than the dilute, as indicated symbolically. For C_8E_5 the concentrated phase has slightly smaller density at higher temperature than the dilute. Note that after complete phase separation, clouding occurs anew when the temperature is changed, to establish further changes in concentration differences between the two phases. Notation of phases according to Table 2.

because two micelles farther apart in space may interact with each other through the micelles intermediate between them (Ornstein-Zernike-equation). The size of the clusters (known as the correlation length in the solution) increases with increasing temperature, which can be interpreted as an increase of the attractive potential strength. At the cloud point, the correlation length (but not the size of individual micelles) becomes macroscopically large and the solution separates into a dense micellar phase (the former clusters) and one which contains very few micelles (the former regions of depressed concentration). This does not imply that the strength of the attractive potential becomes infinite at the cloud point. If this were so, the micelles would flocculate and precipitate out of solution (to form the unstructured pure detergent phase). Instead, the potential remains finite at that point, and all that happens is that the micelles are rearranged in space, such that two macroscopic regions can be distinguished, one in which they occur densely, and one where they are scarce. Both are micellar phases again; due to subtle density differences with detergent concentration, the phases are separated by height in the test tube. Refractive index differences make them discernible by a well-developed meniscus. Further increase of temperature leads to a further rearrangement; clouding occurs again and the meniscus develops at a different place. All this is apparent from Figure 4 and the construct shown there to estimate the concentration of either phase.

The pictorial description given above can also be formulated in a different way. Onset of attractive interactions with increasing temperature suggests that the phase separation is driven by entropy. (This is borne out by calorimetry; the enthalpic contribution is virtually zero.) It follows that the forces between micelles are mediated by the interaction of the latter with water. Gain of entropy by release of bound water provides the energy for the phase separation. This is consistent with the description given above. Dehydration of the ethylene oxide moiety with increasing temperature has both been verified with neutron scattering[26] and nmr.[30] Intuitively one would expect the cloud point to lie higher in temperature when the head group length is increased (because of increased amounts of hydration water). This is qualitatively borne out (Table 1).

E. SALT EFFECTS

At first sight one would expect aggregation properties of nonionic detergents not to change when the water contains electrolyte; there are no free charges and therefore no direct Coulomb interactions with which ions may interfere. However, salt ions do interfere with the water structure, the tetrahedral network that water forms in bulk.[10,31] This leads to substantial effects, which have been investigated extensively in the literature.[7,32] In summary, the results are as follows:

The CMC of ionic detergents decreases dramatically with added electrolyte.[33] 0.1 M NaCl may lower the CMC by as much as an order of magnitude. Generally, ionics have higher CMC's than nonionics, reflecting the additional electrical work in forming micelles (Coulomb repulsion in the head group palisades). The CMC of nonionic detergents is only slightly lowered by electrolyte,[32,34] whereas urea with its disruptive effect on water structure increases it. Note *en passant* that also methanol and ethanol increase slightly the CMC of nonionics, whereas higher alcohols lower it. Also temperature affects the CMC of these detergents not substantially. However, cloud point temperatures are very sensitive to salts and other additives, and this will be discussed in detail.

We have seen above that micellar interactions leading to the consolution phenomenon are mediated by bound water in the head group region. Similarly it can be surmised that mesophase structures are stabilized by solute/water interactions. We therefore expect that the phase boundaries (e.g., in Figure 1) shift in proportion to added salts. The effect on the consolution boundary consists, to a first approximation, in a shift of the boundary parallel to the temperataure axis (shape changes are discussed in Reference 35). Both cases may occur; the boundary may be shifted to higher or lower temperatures depending on the salt species. This is shown in Figure 5, where the cloud point shifts of a 2% solution of C_8E_5 away from its value observed in pure water (61°C) is shown for several monovalent salts.

Obviously, the shifts are not simply proportional to the ionic strength of the solvent, but are highly salt specific. From the finding that the boundaries undergo parallel shifts (insert in Figure 5), it follows that the effects are, to a first approximation, also independent of the salt/detergent molar ratio, which implies the absence of strong ion-micelle interactions. The phenomenon must therefore be related to changed micelle-water interactions.

Figure 5 shows that the anions effectively determine the cloud point temperatures. Cations are known to be smaller and bind more hydration water than anions.[36] Therefore the shifts

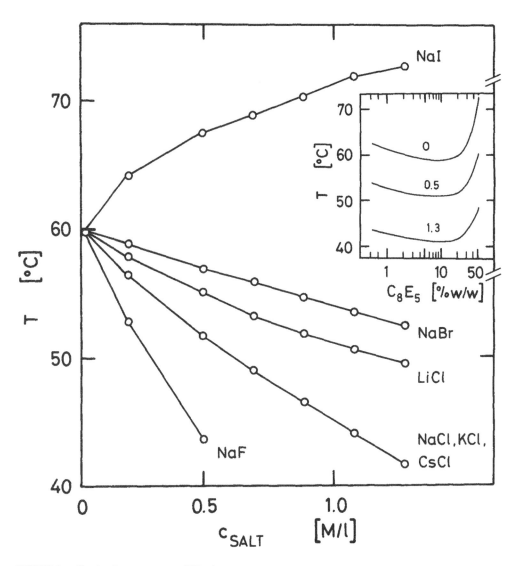

FIGURE 5. Cloud-point temperatures "T" of aqueous solutions of C_8E_5 at 20% w/w as a function of salt molarity (mole/liter) for the salts indicated. The insert shows the consolution boundaries for three concentrations of NaCl: 0, 0.5, and 1.3 M.

are not caused primarily by competition between micelles and ions for free water, but by a more specific anion effect. This becomes evident when the cloud point temperatures, T_d, are plotted as a function of the ionic surface charge density, Z^2e^2/r^2 (Figure 6). With chloride as the common anion, T_d values do not depend in a marked way on the surface charge density of the cations (stars in Figure 6). However, we find a regular decrease in T_d with decreasing size of the anions when sodium is the common cation (dots).

The shifts in temperature of the consolution boundary can be interpreted in terms of micellar solubility, in analogy to protein solubility. The effectiveness of salts in reducing (or enhancing) solubility follows a certain order known as the Hofmeister or lyotropic series.[37] With NaF, LiCl, NaCl, KCl, CsCl, and NaBr, micelles are salted out, whereas salting-in is found with NaI. The solubility changes can be correlated with the structure perturbing ability of the anions. It is well established that structure breaking increases with increasing anionic radius.[38] Thus large ions like iodide perturb the tetrahedral hydrogen-bonding pattern of water more than smaller monovalent anions. However, this correlation is more qualitative

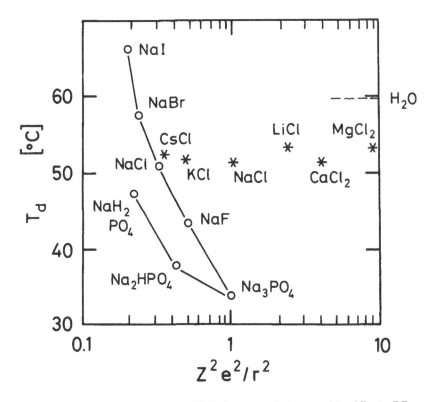

FIGURE 6.   Cloud point temperatures "T_d" of aqueous solutions containing 2% w/w C_8E_5 and 0.5 M salts as a function of surface charge density Z^2e^2/r^2 of either cations (stars) or anions (dots). Z is the charge number, e the electronic charge, and r the crystallographic ion radius.

than quantitative. A more satisfactory description of the solubility changes might involve direct interactions of the ions with the micellar head groups which, in turn, tune micelle-solvent interactions. The picture might be as follows. The strongly hydrated cations are known to be strongly repelled from a dielectric discontinuity, e.g., from a water/air or water/oil interface, by dielectric image-charge forces.[39-41] Anions, being less hydrated, also show less repulsion, with the repulsion decreasing in the order F^-, Cl^-, Br^-, and I^- (iodine may even show a net attraction). Thus, in the vicinity of a micelle surface, a region may exist with decreased cation and anion concentration, for the latter according to the above mentioned series. The corresponding chemical potential difference must then be balanced by changes in the oxyethylene-water interaction, e.g., by subtle changes of the oxyethylene conformation.

The effect of urea and salts with cations capable of forming complexes with ether, (they increase the cloud point temperature), have been reported by Schott and Han.[42]

F. INDUCING PHASE SEPARATIONS BY PRECIPITATING AGENTS

The consolution phenomenon discussed in Section II.D is a particular feature of detergents having oxyethylene head groups (Table 1). It is not commonly found with other detergents, but the exception is hydroxyethanesulfoxide (see Reference 43). However, when precipitating agents used in protein crystallization (e.g., ammonium sulfate, polyethylene glycol, etc.) are added to a micellar solution, one often observes that the solution turns turbid and phase separates after some time in a similar way as described for the polyoxyethylenes. To my knowledge, systematic work to elucidate the mechanism of these phase separations is lacking. Since the phenomenon is important in membrane protein crystallization, a qualitative description is presented here of what can be observed in a special case.

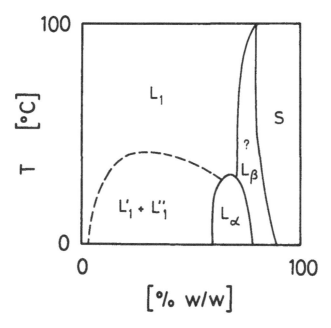

FIGURE 7. Phase diagram of octyl glucoside. Mesophase structures are uncertain. In water alone, a uniform micellar phase L_1 develops at all temperatures. With 10% PEG 4000, a consolution phenomenon occurs whose boundary is shown by dotted curve.

The special case is beta-octyl glucopyranoside (β-OG), precipitated by polyethyleneglycol, e.g., PEG 4000. The detergent dissolves readily in water to form an extended micellar phase. At high concentrations, mesophases occur, whose nature have not been established with certainty (the putative phase diagram is given in Figure 7). Micelles of β-OG are spherical (R ≈ 20 Å) and show, apart from excluded volume effects, no particular interactions in the full temperature interval 0 to 100°C. However, when PEG 4000 is added a critical phenomenon occurs; an isotropic micellar phase prepared at elevated temperatures becomes suddenly turbid when the temperature is lowered below a certain value, and phase separates into two phases, one of which contains most of the detergent and the other most of the PEG plus a few micelles. Preliminary analysis indicates that the detergent rich phase is micellar. The cloud point temperature increases with increasing PEG concentration; the corresponding consolution boundary is shown as a dotted line in Figure 7 for 10% PEG 4000. Note that the mesophase boundaries are not affected appreciably by the presence of PEG, possibly because PEG cannot penetrate into the mesophase structures.

This finding suggests that β-OG-micelles develop attractive interactions in the presence of PEG whose strength increase with decreasing temperature. A possible mechanism for this might be as follows.[44] When two micelles collide as a result of Brownian motion, a region is temporarily created near the faces where they approach that necessarily excludes the polymer. As a consequence, PEG in the adjacent regions will exercise an osmotic pressure which drives the micelles into prolonged contact. This becomes manifested as an apparent attractive potential between the micelles, and leads to concentration fluctuations ("clustering" of the micelles) as described above in Section II.D. The effect is enhanced at low temperature because the polymer has less thermal energy to penetrate into the depleted region.

Similar phase separations can also be induced by ammonium sulfate, e.g. *N,N*-dimethyl odecylamine-*N*-oxide as detergent. Again, the phase separation occurs upon lowering the temperature.

G. WHAT DOES ALL THIS HAVE TO DO WITH MEMBRANE PROTEIN CRYSTALLIZATION?

To answer this question, the literature on membrane protein crystallization must be analyzed and scrutinized for phenomena which can be explained by the concepts of colloid chemistry discussed above. In this book, the leading authors in the field will do it in appropriate detail. It suffices for me to present a brief sketch of the overlap zones where membrane protein and colloid chemistry touch.

The first step towards membrane protein crystallization consists in solubilizing the protein with appropriate detergents, as described in the introduction. Obviously, we can only guess, in general, how the mixed protein-detergent complex looks like. The guess may be guided by electron microscope information from the membrane from where the protein stems. Further information may be obtained from small angle scattering experiments, in particular neutron scattering experiments. There, the contrast (i.e., the power, with which the protein and the detergent scatter neutrons) can be varied, and the two moieties may be localized with respect to each other (at perhaps 30 Å resolution). In any case, the starting information which can guide us in speculating what happens in crystallization assays is quite coarse. Depending on the protein, mixed micelles look like somewhat unusual detergent aggregates from which more or less "naked" protein protrudes. When the protein part is large compared to the detergent coated part, the compexes may essentially behave as soluble proteins, and colloidal aspects become crucial only when, upon assembly in crystals, the detergent coats come into close contact. The detergent may then play a structural role in crystal formation. Detergents with an ease to form mesophases may be less advantageous than others (such as octylglucoside). Section II.B may be useful to help the experimenter identify the mesophases of the detergent he wants to use.

Membrane spanning proteins that do not exceed the bilayer by much on either side of the membrane will possibly form large mixed micelles with substantial parts of their envelope defined by the detergent coat. Small amphiphilic additives may help to smooth out irregularities in the underlying protein surface as pointed out by Michel.[45] When adding precipitating agents to the solutions, the complexes will be expected to behave essentially like micelles. With enough agents and appropriate temperatures, phase separations as described above may occur. The conditions where this happens are specific for the detergent, not the protein, i.e., when one detergent is exchanged for another, crystallization conditions may change despite the protein to be crystallized is the same.

A good example for such a protein is matrix porin, whose dimensions are discussed in Garavito's contribution. When the first crystals of porin were grown,[46] it was noticed that they appear in "oily droplets". *A posteriori* it became clear that these droplets appear in vapor diffusion experiments as a consequence of the phase separations discussed in Section II.D. To describe this, let me consider the special case of octyltetraoxyethylenemonoether (C_8E_4), in which crystals of porin can actually be obtained. A 1% solution of C_8E_4 (roughly 5 times the CMC) has a cloud point at 46°C. In vapor diffusion experiments, appropriate quantities of PEG and salt are added to the solubilizing solutions and brought into vapor contact with aqueous solutions containing higher amounts of PEG and salt, such that water will diffuse from the detergent solution to the reservoir in the course of time. At the start of the vapor diffusion experiment, the cloud point of the detergent solution may be as low as 25 to 30°C. Since water is gradually removed, the PEG and salt concentration increases and the cloud point is lowered further. At ambient temperature, the detergent solution will finally become critical and start to phase separate; gravitational and surface tension effects lead to a detergent-rich phase in the form of small "droplets". Note that in this case the same result is obtained without precipitating agents, but simply by heating the micellar solution slowly above the cloud point.

A glance at Figure 4 shows that the droplets formed form a (dilute) 1% solution of C_8E_4

make up only for a small fraction of the liquid volume, although they contain most of the detergent. It is easy to imagine what happens when the detergent solution contains solubilized protein. As a solution constituent, porin looks somewhat like a large micelle. Upon phase separation, the complexes will be transferred, as most micelles, to the droplets of the detergent-rich phase. Thus, the phase transition brings about a considerable local concentration of the protein in the droplets. There is then possibly a second phase transition in which the protein forms a crystal by (polar?) protein contacts, if the detergent coat allows them to establish. The length of the detergent may be crucial at that step.

Note that the process can be described analogously when C_8E_4 is replaced, e.g., by octylglucoside, with reversed temperature role. With small amounts of precipitating agents, the cloud point may be as low as 10 to 15°C. In vapor diffusion, or by cooling the detergent solution slowly, it will become critical and phase separation with droplet formation occurs.

The protein may not like the high detergent concentrations obtained by phase separation. Bacteriorhodopsin is reported to denature there.[47] The question therefore arises whether protein crystallization is also feasible without the associated detergent phase separation. The answer is yes, but not too far away from the consolution phenomenon. Garavito has obtained the best crystals just before macroscopic droplets form. People working in the field agree that best results are obtained when the crystallization assays are conducted near to the consolution boundary, but without crossing it. These conditions can be found by performing pilot experiments in the absence of protein.

REFERENCES

1. **Helenius, A. and Simons, K.,** *Biochim. Biophys. Acta,* 415, 29, 1975.
2. **Helenius, A., Darrell, R., McCaslin, R., Fries, E., and Tanford, C.,** *Methods Enzymol.,* 56, 734, 1979.
3. **Reynolds, J. A.,** in *Physics of Amphiphiles: Micelles, Vesicles and Microemulsions,* Degiorgio, V., Ed., North-Holland, Amsterdam, 1985, 555.
4. **Wennerström, H. and Lindman, B.,** *Phys. Rep.,* 52, 1, 1979.
5. **Tiddy, G. J. T.,** *Phys. Rep.,* 57, 1, 1980.
6. **Mitchell, D. J., Tiddy, G. J. T., Waring, L., Bostock, T., and McDonald, M. P.,** *J. Chem. Soc., Faraday Trans. 1,* 79, 975, 1981.
7. **Attwood, D. and Florence A. T.,** *Surfactant Systems,* Chapman & Hall, London, 1983.
8. **Small, D. M.,** *J. Am. Oil Chem. Soc.,* 45, 108, 1968.
9. **Tanford, C.,** *The Hydrophobic Effect,* John Wiley & Sons, New York, 1980.
10. **Némethy, G. and Sheraga, H. S.,** *J. Chem. Phys.,* 36, 3382, 1962.
11. **Tanford, C.,** *J. Phys. Chem.,* 76, 3020, 1972.
12. **Dill, K. A.,** in *Physics of Amphiphiles: Micelles, Vesicles and Microemulsions,* Degiorgio, V., Ed., North-Holland, Amsterdam, 1985, 377.
13. **Bendedouch, D. and Chen, S.-H.,** *J. Phys. Chem.,* 87, 153, 1983.
14. **Graber, E., Lang J., and Zana, R.,** *Kolloid Z.,* 238, 470, 1970.
15. **Zana, R. and Lang, J.,** *C. R. Acad. Sci. Sec. C,* 266, 891, 1968.
16. **Gunnarsson, G., Jönson, B., and Wennerström, H.,** *J. Phys. Chem.,* 84, 3114, 1980.
17. **Rosevear, F. B.,** *J. Am. Oil Chem. Soc.,* 31, 628, 1954.
18. **Andersson, B. and Olofsson, G.,** *Colloid Polymer Sci.,* 265, 318, 1987.
19. **Zulauf, M., Fürstenberger, U., Grabo, M., Jäggi, P., Regenass, M., and Rosenbusch, J. P.,** *Methods Enzymol.,* in press.
20. **Shinoda, K.,** in *Colloidal Surfactants,* Shinoda, K., Tamamuski, B. I., Nakagawe, T., and Isemuar, T., Eds., Academic Press, New York, 1963. 1.
21. **Barry, B. W. and El Eini, D. I. D.,** *J. Coll. Interface Sci.,* 54, 339, 1976.

22. **Hayter, J. B. and Zulauf, M.,** *Colloid Polym. Sci.,* 260, 1023, 1982.
23. **Triolo, R., Magid, L. J., Johnson, J. S., and Child, H. R.,** *J. Phys. Chem.,* 86, 3689, 1982.
24. **Zulauf, M. and Rosenbusch, J. P.,** *J. Phys, Chem.,* 87, 856, 1983.
25. **Magid, L. J., Triolo, R., and Johnson, J. S.,** *J. Phys. Chem.,* 88, 5730, 1984.
26. **Zulauf, M., Weckström, K., Hayter, J. B., Degiorgio, V., and Corti, M.,** *J. Phys. Chem.,* 89, 3411, 1985.
27. **Ravey, J. C.,** *J. Colloid Interface Sci.,* 94, 289, 1983.
28. **Brown, W., Johnsen, R., Stilbs, P., and Lindman, B.,** *J. Phys. Chem.,* 87, 4548, 1983.
29. **Strey, R. and Pakusch, A.,** *Proc. 5th Int. Symp. Surfacants in Solution,* Mittal, K. L. and Bothorel, P., Eds., Plenum Press, New York, 1985.
30. **Nilsson, P. G. and Lindman, B.,** *J. Phys. Chem.,* 87, 4756, 1983.
31. **Franks, F.,** *Water,* The Royal Society of Chemistry, London, 1983.
32. **Becher, P.,** *Nonionic Detergents,* Schick, M. J., Ed., Marcel Dekker, New York, 1966, 478.
33. **Reynolds, J.A. and Tanford, C.,** *Proc. Natl. Acad. Sci. USA,* 66, 1002, 1970.
34. **Schott, H.,** *J. Colloid Interface Sci.,* 43, 150, 1973.
35. **Weckström, K. and Zulauf, M.,** *J. Chem.. Soc., Faraday Trans. 1,* 81, 2947, 1985.
36. **Samoilov, O. Ya.,** *Discuss. Faraday Soc.,* 24, 141, 1957.
37. **Hofmeister, F.,** *Arch. Exp. Pathol. Pharmakol.* 24, 247, 1888.
38. **Verrall, R. E.,** *Water, A Comprehensive Treatise,* Vol. 3, Franks, F., Ed., Plenum Press, New York, 1973, 211.
39. **Aveyard, R. and Saleem, S. M.,** *J.Chem. Soc., Faraday Trans. 1,* 72, 1609, 1976.
40. **Aveyard, R., Saleem, S. M. and Heselden, R.,** *J. Chem. Soc., Faraday Trans. 1,* 73, 84, 1977.
41. **Convay, B. E.,** *Adv. Colloid Interface Sci.,* 8, 91, 1977.
42. **Schott, H. and Han, S. K.,** *J. Pharm, Sci.,* 64, 658, 1975, and 65, 975, 1976.
43. **Garavito, R. M. and Rosenbusch, J. P.,** *Methods Enzymol.,* 125, 309, 1986.
44. **Gast, A. P., Hall, C. K., and Russel, W. B.,** *Faraday Discuss. Chem. Soc.,* 76, 224, 1983.
45. **Michel, J.,** *Trends Biochem, Sci.,* 8, 56, 1983.
46. **Garavito, R. M. and Rosenbusch, J. P.,** *J. Cell Biol.,* 86, 327, 1980.
47. **Michel, H.,** *EMBO J.,* 1, 1267, 1982.
48. **Kratohvil, J. P., Hsu, W. P., Jacobs, M. A., Aminabhari, T. M., and Mukunoki, Y.,** *Colloid, Polymer Sci.,* 261, 781, 1983.

Chapter 3

GENERAL AND PRACTICAL ASPECTS OF MEMBRANE PROTEIN CRYSTALLIZATION

Hartmut Michel

TABLE OF CONTENTS

I. INTRODUCTION

At present the structure of more than 400 different proteins are known at or nearly at atomic resolution; only two of them are membrane proteins or complexes of membrane proteins. The striking difference is due to the peculiar properties of membrane proteins, which in turn are a corollary of their location in biological membranes. Biological membranes are made up of bilayers of lipids. The polar head groups of the lipids face the aqueous environment on both sides of the membrane, whereas their hydrophobic alkane chains form the electrically insulating interior of the bilayer. Pictorially speaking, membrane proteins have their heads and feet in the water, but their body is surrounded by a viscous, oily liquid. Thus, in accord with their location, membrane proteins have two polar surface domains on either side of the membrane. These two polar surface domains are separated by a continuous, 30 Å wide belt-like hydrophobic surface domain. As a result membrane proteins are not soluble in ordinary aqueous buffers. For solubilization, one has to add detergents. When used at concentrations above their critical micellar concentration (CMC) the detergents are bound to the hydrophobic surface of the membrane protein or membrane protein complexes via their hydrophobic parts primarily via hydrophobic interactions. The most critical point in handling the membrane proteins is to find a detergent or detergent mixture which *perfectly* replaces the lipids of the membrane. "Perfect" means that there is no decrease in stability of the membrane protein and no change in its structure and function.

II. CLASSIFICATION OF MEMBRANE BOUND PROTEINS, MEMBRANE PROTEINS, AND MEMBRANE PROTEIN COMPLEXES

A substantial number of membrane bound proteins are attached to biological membranes by polar interactions only. These proteins can be released to the aqueous milieu frequently by washes with high salt, EDTA, or alkaline solutions. Treatment with 0.1 M sodium hydroxide or 1 M potassium carbonate has widely been used to discriminate between proteins attached to membranes and real "integral membrane proteins". If a protein can be extracted from the membrane by these agents it is considered to be attached to the membrane, and not to be an integral membrane protein. In a similar way an integral membrane protein is operationally defined by its insolubility in the absence of detergents.

Within the integral membrane proteins one has to discriminate between membrane proteins which are anchored to membranes, but the majority of the protein protrudes either on the outer or inner side of the membrane, and "true" membrane proteins which are primarily located in the membrane. Most frequently the membrane anchor is formed by a single membrane spanning helix. This is the case for the so-called H subunits of the photosynthetic reaction centers from the purple bacteria, where the structures have been determined by X-ray crystallography.[1-3] The protein mass is oriented towards the inner, cytoplasmic side of the membrane. Many surface proteins which face the extracellular milieu of eukaryotic cells possess this type of membrane anchor. The most promising approach to elucidate the structure of these proteins is to clip off the membrane anchor by proteases and to crystallize the soluble part of the protein like an ordinary soluble protein.

There are already four examples where the crystal structures of the water-soluble domains have been determined at high resolution: cytochrome b$_5$,[4] hemagglutinin[5] and neuraminidase[6] from influenza virus, and the human class I histocompatibility antigen HLA-A2.[7] This route should be chosen if possible, since it has a much higher chance of success.

There is one report about the crystallization of a membrane-anchored protein.[8] However, the crystals of the hydrophilic domain alone, which could also be obtained are of a much better quality.

Fatty acids also can serve to anchor proteins to membranes. In the operational definition used above these proteins belong to the integral membrane proteins. The bacterial lipoproteins possess a diglyceride which is bound to an N-terminal cysteine via a thioether bond (see e.g., References 9 to 11). The cytochrome subunit of the photosynthetic reaction center from the purple bacterium *Rhodopseudomonas (Rps.) viridis* belongs to this class. However, the lipid anchor is not visible in the electron density map due to disorder in the detergent micelle which surrounds the hydrophobic part of the reaction center.[13] In eukaryotic membrane proteins myristoylation of the N-terminus, palmitoylation of cysteines (see References 14 and 15 and references therein) as well as covalent attachment of phosphatidylinositol are found (for review see Reference 16). The true membrane proteins are those with multiple membrane spanning segments. The classical example is bacteriorhodopsin, where Henderson and Unwin have provided evidence for the presence of seven membrane-spanning helices.[17] The existence of five membrane-spanning helices as the only transmembrane elements in the L- and M-subunits of the photosynthetic reaction centers from *Rps. viridis*[1,18] and *Rhodobacter (Rb.) sphaerodes*[2] is now firmly proven by X-ray crystallography. Hydrophobic stretches of about 20 amino acids can be found in the sequences of many membrane proteins. These are generally considered to be indicative of membrane spanning helices which are expected to be a general feature of membrane proteins. The well established exceptions are the proteins from the outer membranes of gram-negative bacateria called omps or porins. Spectroscopic and crystallographic results indicate that these proteins contain β-sheets rather than α-helices as membrane spanning segments.[19-23] This structural difference might be a consequence of the location of these proteins in the outer membrane, since it is known that long stretches of hydrophobic amino acids which are absent in the porins act as signals to stop the transfer of proteins through the inner bacterial membrane. Therefore proteins with such hydrophobic stretches might be unable to cross the inner membrane of bacteria and to reach their destination in the outer membrane.

III. PURIFICATION OF MEMBRANE PROTEINS

Although purity, homogeneity, and stability of membrane proteins are the most important prerequisites for their crystallization, we can only discuss these aspects briefly. For extended discussions on the purification of membranes and membrane proteins we refer to *Methods in Enzymology*. Twenty volumes of this series are devoted to studies on membrane proteins. Excellent introductions to membrane and membrane protein purifications are available.[24,25]

A. ISOLATION OF MEMBRANES

The most suited starting materials for membrane protein isolation are specialized biological membranes, where certain membrane proteins are considerably enriched. Typical examples are photosynthetic membranes of various photosynthetic organisms or organelles, the membranes from electric organs, the sarcoplasmic reticulum membrane, and the disc membranes from rod outer segments. The halobacterial purple membrane contains only one protein component.[26] The simpler the composition of the membrane the better it is suited for large scale isolation and purification of membrane proteins as needed for continuous crystallization attempts. Plasma membrane proteins of low abundancy have to be considered as unfavorable, but challenging choices. Membrane protein complexes like photosynthetic reaction centers, cytochrome oxidases, and the bc_1-complexes isolated from bacteria frequently possess a much simpler composition than their counterparts from higher organisms. It is thus frequently easier to purify them to homogeneity.

The first step is to break the cells, frequently by mechanical treatments using homogenizers, the "French press", gas cavitation (e.g., "Yeda press"), sonication or osmotic lysis, sometimes after enzymatic digestions. The membranes and organelles are separated from the undisrupted cells by differential centrifugation and further purified by isopycnic centrifugation in the ultracentrifuge using sucrose, Percoll®, or Metrizamide® to establish the density gradients. The differences in the density of the various membranes are mainly based on their different protein/lipid ratios and their contents of other pigments. Use of Percoll® has the advantage that isoosmotic conditions can be maintained. Purified organelles can then be ruptured by osmotic lysis. Another recommendable procedure is aqueous two-phase separation using dextran/polyethyleneglycol.[27] This procedure takes advantage of different surface properties of membranes. Therefore, even inside-out and inside-in vesicles of the same membrane can be separated.

B. ISOLATION OF MEMBRANE PROTEINS

In several instances it proved to be of advantage to remove peripheral membrane proteins first by washes with high salt and/or EDTA and/or a high pH treatment. However, these treatments might also remove peripheral membrane proteins which are bound to integral membrane proteins in their functional state. High pH treatment may also cause some denaturation of the integral membrane proteins.

The only way to solubilize membrane proteins with large hydrophilic and hydrophobic domains is to add detergents and to cover the hydrophobic domains of the membrane protein with a layer of detergent. It is frequently possible to extract specifically one or the other membrane protein at low detergent-to-protein ratios. Protocols for the routine isolation of the photosynthetic reaction center from the purple bacterium *Rhodobacter sphaeroides*[28] and the complexes of reaction centers with the inner light-harvesting antenna from other purple bacteria (Michel, unpublished) are based on this method. It critically depends on the choice of pH, ionic strength, and temperature.

The inverse strategy can also be used to purify membrane proteins. One can remove all but one membrane protein complex from the membrane by subsequent detergent extractions. The classical example is cytochrome oxidase from bovine heart mitochondria, which even forms two-dimensional crystals spontaneously after this treatment.[29]

The generally used strategy to purify membrane proteins is a ''brute force'' strategy: one tries to solubilize the membrane completely by adding a vast excess of detergents and to separate its individual components in their detergent micelles afterwards. In an operational definition a membrane protein is considered ''solubilized'' when it is found in the supernatant after 1 h of centrifugation at 100,000 g. In this sense complete solubilization is indicated by the absence of any pellet after 1 h centrifugation at 100,000 g. One critical parameter is the detergent-to-lipid ratio. A situation would be optimal, where only one lipid molecule would be incorporated in one detergent micelle, and lipid-free individual membrane proteins or protein complexes would be found in separate detergent micelles. We can only approach this situation when we have an excess of pure detergent micelles present at the same time. Details of the concepts of membrane protein solubilization are described by Helenius and Simons[30] and Helenius et al.[31]

For the initial solubilization the detergents are selected primarily according to the following criteria: (1) their ability to solubilize that particular membrane, this in turn depends on the complex interactions between membrane proteins, detergents, and lipids; and (2) stability of the membrane proteins to be isolated in the presence of the detergent. Detergents with small polar head groups tend to solubilize better than those with large polar headgroups but they destabilize membrane proteins to a larger extent. It is possible to solubilize the membrane in one detergent, and to use milder detergent for membrane protein purification. Within a homologous series of alkyl detergents, an increase in membrane protein stability is observed with an increase of the length of the alkyl chain. For instance, under suboptimal pH-conditions solubilized bacteriorhodopsin is more stable by roughly a factor of 3 when nonyl-β,D-glucopyranoside is used instead of octyl-β,D-glucopyranoside (Michel, unpublished). With longer alkyl chains, the lower solubility of the alkyl glucopyranosides becomes a problem. In the case of the Ca-ATPase from sarcoplasmic reticulum with other polar head groups a gain in stability was observed up to an alkyl chainlength of twelve, and no further increase in stability was seen with longer alkyl chains.[32]

Among the detergents with an alkyl chain those with charged head groups (dodecyl-sulfate, hexadecyltrimethylammonium) are generally not suited for the isolation of "native" membrane proteins. However, there are exceptions like some photosystem I reaction centers from cyanobacteria.[33] Zwitterionic alkyl-detergents (e. g., sulfobetaines) and those with an N-oxide as polar head group can be used with many more proteins, but they are still too harsh for most membrane proteins. Among the detergents with an alkyl chain, those with polyoxyethylene (e.g., $C_{12}E_8$; Lubrol-series; Triton X-series; Triton N-series; Brij-series) or a disaccharide head group are the mildest. The former have the disadvantage of being susceptible to autoxidation. They can be fully reduced by treatment with sodium borohydride and subsequent passage through a mixed bed ion-exchanger.[34] Dodecyl-β,D-maltoside seems to be ideally suited for many membrane proteins of bacterial or mitochondrial origin. For many membrane proteins of eukaryotic origin the bile salt derivative 3-(3-Cholamidopro-pyl)dimethylammonio-1-propanesulphonate (CHAPS) seems to be an excellent choice.[35]

The separation methods for membrane proteins are essentially the same as for soluble proteins. It is normally sufficient to work at detergent concentrations slightly above the CMC. Frequently *ion-exchange chromatography* which is based on surface charges of the protein is the method of choice, especially in the modern FPLC-version. Both anion and cation exchangers can be used but neither can be used with ionic detergents. Zwitterionic detergents also possess disadvantages due to electrostatic interactions as a consequence of the strong dipole moment at their polar head groups. The column must be equilibrated with the detergent containing buffer prior to its use. The Pharmacia corporation (Uppsala, Sweden) provides a complete reference list upon request. Extremely powerful methods are isoelectric focusing, where the commercially available preparative scale equipment can be used es-pecially in connection with alkylglucosides, alkylthioglucosides, and alkylmaltosides as detergents, and chromatofocusing. In both methods proteins are separated according to their isoelectric points. In the chromatofocusing technique the proteins are bound to special ampholytic column materials and eluted with a focusing pH-gradient. In addition, isoelectric focusing is the best analytical method to test purity and homogeneity of the final membrane protein preparation. Isopycnic sucrose *density gradient centrifugation* is also very useful for the purification of membrane proteins, in contrast to the purification of soluble proteins. The density of the pure proteins is more or less the same. Since we purify mixed protein-detergent micelles, isopycnic ultracentrifugation separates membrane proteins according to their detergent to protein ratio. Pigments, when present, also influence the density. Solu-bilized chlorophyll proteins can be easily purified by this method. Its power, however, has not been fully explored with other membrane proteins.

Molecular sieve chromatography is less powerful for membrane proteins than for soluble proteins, since the detergent-protein complexes are separated as a function of the sum of the molecular weights of bound detergents plus protein. Therefore, it is of advantage to use detergents which form micelles with a small micellar weight. Then the molecular weight of the protein dominates the weight of the complex. Especially suited as detergent seem to be CHAPS (micellar weight: 6150[35]), octyl-β-D-glucopyranoside (micellar weight: 8000[36]) and N,N-dimethyldodecylamine-N-oxide (micellar weight: 16,000[37]) as compared to Triton X-100 (micellar weight: 90,000[30]). Molecular sieve chromatography also removes lipids very efficiently, when stronger detergents are used.

Affinity chromagraphy is of course of great value in the few specific cases where it can be used. A good example is the β-adrenergic receptor.[38]

Less popular are the classical *precipitation methods* using ammonium sulfate or poly-

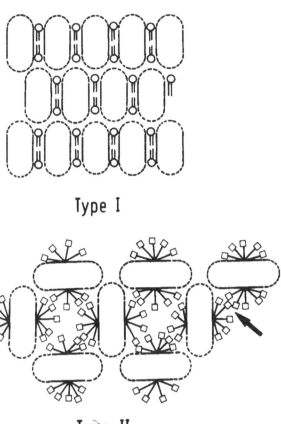

Type I

Type II

FIGURE 1. The two basic types of membrane protein crystals shown as cross-sections. Type I: stacks of membranes containing the two-dimensionally crystalline membrane proteins which are then ordered in the third dimension. Type II: a membrane protein crystallized with detergents bound to its hydrophobic surface. The polar surface parts of membrane proteins are indicated by dashed lines, the hydrophobic surface parts by continuous lines. Lipid molecules are shown as squares with two straight lines attached to the squares symbolising alkyl chains, detergent molecules are shown as circles with one straight line attached to the circle. (Taken from Michel, H., *Trends Biochem. Sci.*, 8, 56, 1983.)

ethylenglycol. They can be tried, since they will also be used in the various crystallization attempts.

IV. CRYSTALLIZATION OF MEMBRANE PROTEINS

A. GENERAL ASPECTS

In principle, membrane proteins can form two different types of three-dimensional crystals.[39] They are reproduced in Figure 1. Three-dimensional crystals of type I are stacks of two-dimensional crystals. The two-dimensional crystals are formed in the plane of the membrane with the lipid bilayer still present. In the third dimension these two-dimensional crystals have to be ordered with respect to rotation, translation and sidedness (up and down orientation). This can be achieved either during or after the formation of the two-dimensional crystals. Actually crystals of this type were the first reproducible true three-dimensional crystals of membrane proteins which were observed.[28] Since type I crystals are held together by polar interactions (perpendicular to the membrane and possibly also in the planes of the membrane along the polar headgroups of the lipids) as well as by hydrophobic interacations (in the planes of the membranes) it would be desirable to increase gradually and simultaneously both kinds of interactions. However, there are no straight forward methods to achieve this. The most promising route is to remove the detergent from solubilized membrane proteins by slow dialysis at elevated ionic strength.[40] It might be advisable to add a small amount of lipids to allow a bilayer to be formed[41] (see also Chapter 11). The prospect of obtaining three dimensional type I crystals which are well-ordered and large enough for an X-ray structure analysis is rather low.

The alternative is to crystallize the membrane proteins with their hydrophobic surface coated by detergents. The crystal lattice is formed by the protein via polar interactions between the polar surface domains of the membrane proteins. The detergent plays a passive, but crucial role. It still surrounds the membrane protein in the same micellar way as in solution. One can also imagine that fusion of micelles occurs in a way that at the end the crystal might contain a continuous monolayer of detergent in the form of a two- or even three-dimensional network.

In this view the driving forces for the crystallization, the polar interactions between the polar surface domains of the membrane proteins, are the same as for the crystallization of soluble proteins. Therefore, type II membrane protein crystals are grown by the same basic techniques as crystals of water soluble proteins (see Chapter 1).

B. THE ROLE OF THE DETERGENT

The difference between the crystals of type I and type II is the presence of detergents in type II crystals. The detergents cause a variety of conceptual and practical problems. These are discussed in the following paragraphs.

1. Role of the Polar Head Group

Since the detergent is bound to the protein's hydrophobic surface primarily via hydrophobic interactions of its alkyl tail, its polar head group necessarily covers a zone of the protein's polar surface (C in Figure 2A). This zone is not available for productive polar interactions which might lead to the formation of crystal nuclei. As a consequence the polar head group of the detergent should be as small as possible. In this sense, the N,N-dimethylamine-N-oxide head group used in the crystallization of the photosynthetic reaction centers from the purple bacterium *Rhodopseudomonas viridis* is close to ideal. Then detergents with hydroxyethylsulfoxide as head group, followed by 2,3-dihydroxypropylsulfoxide and then by β-D-glucopyranoside and β-D-thioglucopyranoside as head groups are most suited for crystallization attempts. Evidently, the size of the polar head group is less important when the extramembraneous protein domains are large. It is therefore no surprise that photosystem I reaction centers can also be crystallized using dodecyl-β-D-maltoside or even Triton X-100 as detergent.[33] Unfortunately, detergents with small headgroups are much more denaturing than those with large head groups.

2. Role of the Alkyl Chain

The length of the alkyl chains is of crucial importance for the diameter of the detergent micelle and therefore for the thickness of the detergent belt as shown in Figure 2B. With thick detergent belts the range of possible contact angles between two membrane proteins is much more narrow, than with thinner detergent belts, due to the increased steric hindrance. However again, what seems to be most suited for the crystallization of membrane proteins is less suited for their stability. As mentioned in the previous section, there is a severe loss of stability when detergents with an octyl-chain are used instead of detergents with longer aklyl chains. There is another reason, why detergents with larger alkyl chains might sometimes be more suited; it is discussed in Section IV.D.

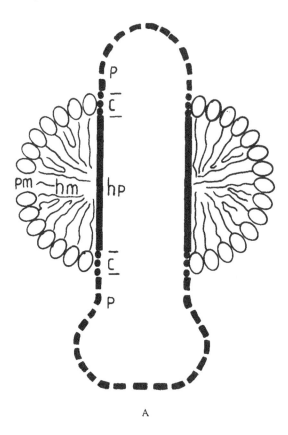

A

FIGURE 2. (A) Cross-section through a protein containing detergent micelle. The hydrophobic surface parts (h) are shown as straight lines, the polar surface parts (p) available for protein-protein interactions as dashed lines, the polar surface part covered (c) by the polar part (pm) of the detergent micelle by dotted lines. The hydrophobic interior of the detergent micelle is indicated by hm. (B) Two membrane proteins interacting at their polar surface domains with their hydrophobic surface covered by a thinner monolayer-like micelle (top) or by a spherical micelle (bottom). The thinner micelle allows the formation of a smaller contact angle x than the spherical micelle.

3. Specific Detergent-Protein Interactions

The refined 2.3 Å resolution structure of the photosynthetic reaction center from *Rhodopseudomonas viridis* revealed electron density only for one detergent molecule (*N,N*-dimethyldodecylamine-*N*-oxide). This detergent molecule is bound into a pocket formed by two protein subunits and pigments (Deisenhofer, Epp, Sinning, Michel, manuscript in preparation). Even polar interactions between the detergent's polar head group and neighboring polar groups of the protein are possible. The pocket is not large enough to accommodate detergents with larger head groups. This fact could explain why other detergents could not be used to crystallize this photosynthetic reaction center.

4. Phase Separation

One of the worst obstacles in membrane protein crystallization is the detergent's tendency to form a separate, viscous detergent phase. This detergent phase is formed upon the addition of precipitants and/or temperature shifts and contains the membrane proteins. This behavior of the detergents is extensively discussed in Zulauf's contribution (Chapter 2). Only a few observations from our laboratory are listed here, which should be helpful to the unexperienced crystal grower. Many membrane proteins undergo a rapid denaturation in the detergent

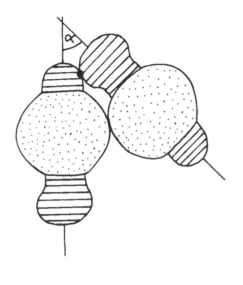

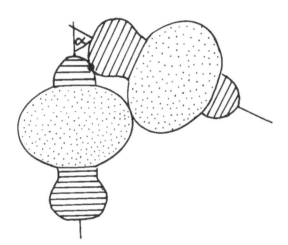

FIGURE 2B.

phase. There are, however, a few examples, where crystals are formed in the detergent phase. In the case of the B800/850 light-harvesting complex from *Rhodospirillum molischianum* crystals can be formed inside and outside of the detergent droplets, but with different morphology (unpublished observations). Of great importance are the countermeasures to shift or to suppress the phase separation. When polyethylenglycol (PEG) is used as precipitating agent, phase separation is more pronounced with increasing molecular weights of the polyethylenglycols (PEG 600 < PEG 1000 < PEG 2000 < PEG 4000 < PEG 6000). When salts are used, phase separation is less pronounced with ammonium sulfate than with potassium or sodium phosphate. The phase separation can be shifted to higher salt concentrations by the addition of, e.g., glycerol or pipecolinic acid. Fortunately variation of temperature has often opposite effects on protein solubility and phase transition (''consolution boundary''). With *N,N*-dimethyldodecylamine-*N*-oxide and octyl-β-D-glucopyranoside, the phase transition shifts to higher precipitant concentrations with increasing temperature, whereas the protein solubility frequently increases with decreasing temperature. Therefore use of higher temperatures (25°C) is recommended in such cases. When phase separation is a problem it is also desirable to work rather close at the protein's isoelectric point where protein solubility is at its minimum. Below a certain temperature ("Krafft point"), many detergents crystallize. Detergent crystals can also be obtained upon addition of the precipitating agents. The crystals may be mistaken as membrane protein crystals. Well known for this behavior at 4°C are octyl-β-D-thioglucoside, decanoyl-N-methylglucamide and dodecyl-β-D-maltoside.

C. THE SMALL AMPHIPHILE CONCEPT

A very successful idea has been the addition of small amphiphilic molecules.[39] The rationale behind it is that these compounds form mixed micelles with the detergent. These mixed micelles are smaller than pure detergent micelles, and might thus be accommodated in the protein's crystal lattice more easily. Some of them might replace detergent molecules at positions critical for the formation of the crystal lattice. The smaller head groups of the small amphiphiles as compared to the detergents' head group cover less of the protein's polar surface and leave more of it available for crystallization. Specific interactions of the small amphiphile with the membrane protein have also to be considered since only the *threo*-form of heptane-1,2,3-triol ("high melting isomer"), but not the *erythro*-form ("low melting isomer") can be used to crystallize the photosynthetic reaction center from *Rps. viridis*. The difference between a small amphiphile and a detergent is that the small amphiphile alone is unable to form micelles.

One of the arguments presented above has now been proven since the addition of 3 to 5% heptane-1,2,3-triol has been shown to reduce the micellar weight of octyl-β-D-glucopyranoside micelles by 50% (Hunt, Kataoka, Fujisawa and Engelmann, personal communication).

About 100 amphiphilic compounds have been used in our attempts to improve the quality of the bacteriorhodopsin crystals. Most of them were either insoluble at the necessarily high salt concentrations, or had denaturing effects under the conditions used. At present we use mainly heptane-1,2,3-triol[39] and benzamidine.[42] Although heptane-1,2,3-triol is the most successful compound so far, it works best in the combination with salts as precipitating agents, whereas benzamidine frequently works better with polyethylenglycol.

An interesting observation concerning the action of the small amphiphiles has been made during our attempts to crystallize the B800/850 light-harvesting complex from *Rhodospirillum molischianum*. One promising crystal form (Figure 3A) is only obtained when heptane-1,2,3-triol and the membrane protein are brought to supersaturation at the same time. Tetragonal crystals of the B800/850 complex are observed first. Later heptane-1,2,3-triol crystallizes and the protein crystals crack (Figures 3B and C) even when the heptane-1,2,3-triol crystals do not touch the protein crystals. Then the B800/850 crystals redissolve completely and finally a different crystal form (Figure 3D) is formed. Most likely, upon the crystallization of heptane-1,2,3-triol the mixed detergent/heptane-1,2,3-triol micelles and the detergent belts surrounding the protein release bound heptane-1,2,3-triol and take up more detergent from the solution. As a result the detergent replaces heptane-1,2,3-triol, the micelles become larger thereby breaking the crystal lattice, which is formed by tightly packed B800/850 complexes.

D. ENERGETICS OF CRYSTAL FORMATION

In the case of the photosynthetic reaction center from *Rps. viridis* the location of the detergent N,N-dimethyldocylamine-N-oxide has been determined by neutron diffraction and H_2O/D_2O contrast variation at low resolution.[13] Part of the results are reproduced in Figure 4. It has been shown that the detergent forms a belt like micelle around the membrane-spanning helices of the reaction center. However, the micelles touch each other. They form "bridges" of detergents. It is not entirely clear at present whether these bridges are points of fusion of micelles, or if they are simple contact sites between micelles without fusion.

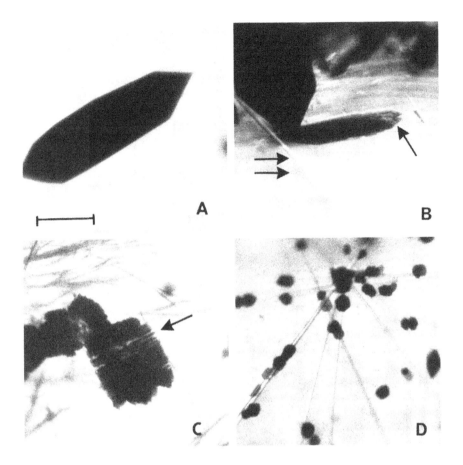

FIGURE 3. Crystals of the B800/850 light-harvesting complex from *Rhodospirillum molischianum* grown at 18°C; starting with 1.8 *M* ammonium sulfate, 0.1% *N,N*-dimethyldodecylamine-*N*-oxide as detergent, 3% heptane-1,2,3-triol, by vapor diffusion against 3.5 *M* ammonium sulfate. Crystals diffracting X-rays to 2.4 Å resolution, space group P42₁2, are obtained first (A). Then transparent heptane-1,2,3-triol crystals (B, double arrow) are formed, the colored crystals of the light-harvesting complex crack (B, arrow) and redissolve (C, arrow), finally, different poorly ordered crystals of the light-harvesting complex (D) are obtained. The bar indicates 0.2 mm.

The fact that the thickness of the bridges is about the same as the thickness of the detergent monolayer can be used as an argument in favor of simple contact sites. For a point of fusion one would expect a diameter for the bridge corresponding to the diameter of a pure detergent micelle, or roughly twice of the detergent monolayer. It has also been shown that the regions of protein-protein contact are free of detergent. However, apart from the detergent-detergent contacts there are also polar contacts of detergent micelles with neighboring reaction centers. One therefore has to assume that protein-protein, detergent-detergent, and (polar)protein-detergent contacts all contribute to the energetics of crystal formation and crystal packing. This result readily explains the observation that the photosynthetic reaction center from *Rhodopseudomonas viridis* could not be crystallized when *N,N*-dimethyldodecylamine-*N*-oxide was replaced by *N,N*-dimethyldecylamine-*N*-oxide. Even replacement of the former detergent by the latter in the crystals by soaking lead to disorder. The latter detergent might be too short for the detergent-detergent and (polar)protein-detergent contacts observed with the former detergent. Therefore, the detergent with a shorter alkyl chain is not always more suited for crystallization of membrane proteins than the detergent with a longer chain. This observation also shows that it is important to try to crystallize membrane proteins at conditions

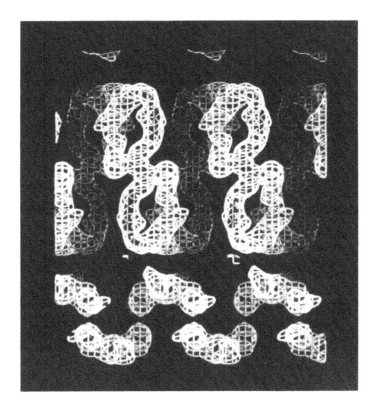

FIGURE 4. The packing of the detergent micelles of the photosynthetic reaction centers from *Rps. viridis* in the P4$_3$2$_1$2 lattice. It can be described as stacking of pairs of ring-like micelles in two rows per unit all parallel to [0,0,1]. The detergent rings of the molecules within a single row are connected two by two. The contacts or "bridges" have a similar cross-section as the rings themselves. These contact sites are located on crystallographic twofold axes and are of two types, a straight one and hairpin one. The detergent is thus distributed in parallel zig-zag chains of interconnected rings throughout the crystal. These chains are parallel to the c-axis and only very loosely connected between them through weak detergent density found in the vicinity of the 4$_3$ crystal axis. View (down 110) of several unit cells showing the packing of the detergent rings.[13]

close to the consolution boundary, where interactions between detergent micelles are enhanced, as first stressed by Garavito et al.[43] (see also Chapters 2 and 4 for discussion).

With most other membrane protein crystals the situation is different. In the case of bacteriorhodopsin, the reaction center from *Rb. sphaeroides*, several light-harvesting complexes from photosynthetic bacteria, and porin from *E. coli* with the same precipitating agent and under the same conditions of ionic strength, pH, and temperature, the same crystal form is obtained when a slightly different detergent is used. These observations probably indicate that protein-protein interactions are the dominant driving forces for the crystallization of these proteins.

E. PRACTICAL ASPECTS

1. Detergent Exchange

Considering the discussions above the choice of the detergent is one of the most critical points in the crystallization of membrane proteins. Normally membrane proteins are isolated using a rather mild, inexpensive detergent like Triton X-100 which is not suited for the crystallization. The first task therefore is to exchange the detergent. This is done most easily by absorbing the membrane protein to a small ion exchange (DEAE, monoQ, CM, SP-column) or hydroxyapatite column and to wash it with several column volumes of buffer containing the new detergent at a concentration higher than the CMC. The membrane protein is then eluted with the new detergent by increasing the ionic strength or by changing the pH. At the same time the protein can be concentrated. The amount of the "old" detergent, which is still present after exchange should be determined. This is most conveniently done by use of radioactively labeled detergents. It will be found that longer washing is needed to exchange the detergent than expected. This is most likely due to the frequent replacement of a detergent with low critical micellar concentration by one with a high critical micellar concentration. A list of the available radioactively labeled detergents can be found in Appendix II at the end of this book. Other convenient methods to exchange the detergent are affinity chromatography, molecular sieve chromatography, or sucrose gradient centrifugation. Extraction of the detergent with organic solvents is not recommended due to denaturation of the membrane proteins.

2. Stability Tests

Another crucial point is to test the stability of the membrane protein to be crystallized under the various conditions. Most important is the pH-dependence of stability. It is rarely found that a membrane protein is most stable at pH 7 or 7.4 which is used by many membrane biochemists for isolation. Our experience shows a pH-optimum of stability frequently at pH 5 to 6 or 8 to 9. The broadness of the pH-optimum of stability depends on the nature of the detergent used. Longer alkyl chains and larger head groups lead to broader pH-optima. If no simple functional or spectroscopic test is available for the determination of protein stability, the most convenient (but sometimes misleading) procedure is to look for the development of a denatured precipitate indicated by an increased turbidity of the protein sample. One should also inspect the turbid protein solution with the light microscope in order to exclude that the turbidity is due to phase separation. A protein concentration of 3 mg/ml is more than sufficient for this simple test.

3. Precipitation of Membrane Proteins

In principle the same precipitants and precipitation methods as for soluble proteins can be used (see Chapter 1). For membrane proteins various polyethylenglycols (mol wt 600 to 8000), ammonium sulfate and sodium or potassium phosphate have been used and good quality crystals of membrane proteins have been obtained with these precipitants. The stability of the membrane protein may also depend on the nature and purity of the precipitant.

Among the various techniques to achieve supersaturation osmotic equilibration of the protein solution with a larger reservoir of higher osmotic pressure through the gas phase ("vapor diffusion") is most convenient. "Hanging" or "sitting drop" methods can both be used, although the hanging drop method is hindered by the fact that the surface tension of the protein solution is decreased due to the presence of detergents and the volume of the droplet therefore has to be reduced.

Crystallization attempts using the dialysis technique are more complicated to set up. The advantage is that the conditions can be more accurately controlled compared to the vapor diffusion technique. Whereas with vapor diffusion salt, protein, detergent, additives are concentrated simultaneously and the pH may change, all these parameters can be varied individually when the dialysis technique is applied.

The speed used to achieve supersaturation should also be varied. During crystallization attempts the membrane protein can undergo several competing reactions. These are indicated in Figure 5. It may be denatured, which is an irreversible process (i). It may form an amorphous precipitate without loss of activity and without apparent denaturation (ii). This is a reversible process. We hope that the membrane protein crystallizes (iii). This again is

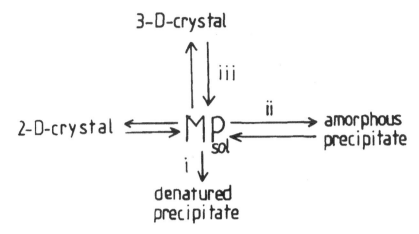

FIGURE 5.   Schematic drawing showing the competing reactions possible for solubilized membrane proteins (MPsol). Denaturation (i) is an irreversible process, crystallization (iii) and formation of an amorphous precipitate (ii) are reversible processes.

a reversible process. Long term experiments favor process (i), whereas fast experiments (rapid supersaturation) favor process (ii). Crystals are frequently obtained only on the intermediate time-scale.

V. CONCLUSIONS

Since the first promising reports on the crystallization of membrane proteins[44,45] many reports have appeared describing the crystallization of membrane proteins including the well-diffracting reaction center crystals from *Rps. viridis*[46] and *Rb. sphaeroides*.[47,48] However, most crystals are too small or too disordered for high-resolution X-ray work. In our laboratory alone about 50 different crystal forms have been obtained from various bacterial light-harvesting complexes and about 10 other membrane proteins. So far only the crystals of bacteriorhodopsin,[42] porin, and maltoporin from *Escherichia coli*[19,49] mammalian prostaglandin H synthase,[50] porin from *Rhodobacter capsulatus*,[51] photosystem I from a thermophilic cyanobacterium[52] and the bacterial B800/850 light-harvesting complexes from *Rps. acidophila*[53] and *Rhodospirillum molischianum* (Michel, to be published, see also Figure 3) might be good enough to yield high resolution structures in addition to the two known reaction center structures.[1,2,3] The way from a first small crystal to a large well-diffracting crystal is usually long, painful, and frustrating. A long term commitment from both the scientist and the funding agency is needed.

ACKNOWLEDGMENT

I highly appreciate the generosity and support by Dieter Oesterhelt, who made our success possible, and last but not least I thank my wife Ilona Leger-Michel for her patience and tolerance.

Financial support was obtained from the Deutsche Forschungsgemeinschaft (SFB143, Leibniz-Programm), the Max-Planck-Gesckschaft and the Fonds der Chemischen Industrie.

REFERENCES

1. **Deisenhofer, J. and Michel, H.,** The photosynthetic reaction centre from the purple bacterium *Rhodopseudomonas viridis, EMBO J.,* 8, 2149, 1989, (Nobel lecture).

2. **Allen, J. P., Feher, G., Yeates, T. O., Komiya, H., and Rees, D. C.,** Structure of the reaction center from Rhodobacter sphaeroides R-26: the protein subunits, *Proc. Natl. Acad. Sci. USA,* 84, 6162, 1987.

3. **Chang, C.-H., Tiede, D., Tang, J., Smith, U., Norris, J. and Schiffer, M.,** Structure of Rhodopseudomonas sphaeroides R-26 reaction center, *FEBS Lett.,* 205, 82, 1986.

4. **Mathews, F. S., Argos, P., and Levine, M.,** The structure of cytochrome b_5 at 2.0 Å resolution, *Cold Spring Harbor Symp. Quant. Biol.,* 36, 387, 1972.

5. **Wilson, I. A., Skehel, J. J., and Wiley, D. C.,** Structure of the haemagglutinin membrane glycoprotein of influenza virus at 3 Å resolution, *Nature,* 289, 366, 1981.

6. **Varghese, J. N., Laver, W. G., and Colman, P. M.,** Structure of the influenza virus glycoprotein antigen neuraminidase at 2.9 Å resolution, *Nature,* 303, 35, 1983.

7. **Bjorkman, P. J., Saper, M. A., Samraoui, B., Bennett, W. S., Strominger, J. L., and Wiley, D. C.,** Structure of the human class I histocompatibility antigen, HLA-A2, *Nature,* 329, 506, 1987.

8. **Miki, K., Kaida, S., Kasai, N., Iyanagi, T., Kobayashi, K., and Hayashi, K.,** Crystallization and Preliminary X-ray Crystallographic Study of NADH-cytochrome b_5 Reductase from Pig Liver Microsomes, *J. Biol. Chem.,* 262, 11801, 1987.

9. **Hantke, K. and Braun, V.,** Covalent Binding of Lipid to Protein, *Eur. J. Biochem.,* 34, 284, 1973.

10. **Pugsley, A. P., Chapon, C., and Schwartz, M.,** Extracellular Pullulanase of *Klebsiella pneumoniae* is a Lipoprotein, *J. Bacteriol.,* 166, 1083, 1986.

11. **Yu, J., Inouye, S., and Inouye, M.,** Lipoprotein-28, a Cytoplasmic Membrane Lipoprotein from *Escherichia coli, J. Biol. Chem.,* 261, 2284, 1986.

12. **Weyer, K. A., Schäfer, W., Lottspeich, F., and Michel, H.,** The Cytochrome Subunit of the Photosynthetic Reaction Center from Rhodopseudomonas viridis Is a Lipoprotein, *Biochemistry,* 26, 2909, 1987.

13. **Roth, M., Lewit-Bentley, A., Michel, H., Deisenhofer, J., Huber, R., and Oesterhelt, D.,** Detergent structure in crystals of a bacterial photosynthetic reaction centre, *Nature,* 340, 659, 1989.

14. **Magee, A. I. and Courtneidge, S. A.,** Two classes of fatty acid acylated proteins exist in eukaryotic cells, *EMBO J.,* 4, 1137, 1985.

15. **Schultz, A. M., Henderson, L. E., and Oroszlan, S.,** Fatty Acylation of Proteins, *Ann. Rev. Cell Biol.,* 4, 611, 1988.

16. **Low, M. G., Ferguson, M. A. J., Futerman, A. H ., and Silman, I.,** Cell-surface anchoring of proteins via glycosylphosphatidylinositol structures, *Ann. Rev. Biochem.,* 57, 285, 1988.

17. **Henderson, R. and Unwin, P. N. T.,** Three-dimensional model of purple membrane obtained by electron microscopy, *Nature,* 257, 28, 1975.

18. **Deisenhofer, J., Epp, O., Miki, K., Huber, R., and Michel, H.,** Structure of the protein subunits in the photosynthetic reaction centre of *Rhodopseudomonas viridis* at 3 Å resolution, Nature, 318, 618, 1985.

19. **Garavito, R. M., Jenkins, J., Jansonius, J. N., Karlsson, R., and Rosenbusch, J. P.,** X-ray diffraction analysis of matrix porin, an integral membrane protein from *Escherichia coli* outer membranes, *J. Mol. Biol.,* 164, 313, 1983.

20. **Kleffel, B., Garavito, R. M., Baumeister, W., and Rosenbusch, J. P.,** Secondary structure of a channel-forming protein: porin from *E. coli* outer membranes, *EMBO J.,* 4, 1589, 1985.

21. **Jap, B. K.,** Molecular design of PhoE porin and its functional consequences, *J. Mol. Biol.,* 205, 407, 1989.

22. **Sass, H. J., Büldt, G., Beckmann, E., Zemlin, F., van Heel, M., Zeitler, E., Rosenbusch, J. P., Dorset, D. L., and Massalski, A.,** Densely packed β-structure at the protein-lipid interface of porin is revealed by high resolution cryo-electron microscopy, *J. Mol. Biol.,* 209, 171, 1989.

23. **Weiss, M. S., Wacker, T., Nestel, U., Woitzik, D., Weckesser, J., Kreutz, W., Welte, W., and Schultz, G. E.,** The structure of porin from *Rhodobacter capsulatus* at 0.6 nm resolution, *FEBS Lett.,* 256, 143, 1989.

24. **Findlay, J. B. C. and Evans, W. H.,** *Biological Membranes - A Practical Approach,* IRL Press, Oxford, 1987.

25. **Datta, D. B.,** A Comprehensive Introduction to Membrane Biochemistry, Floral Publishing, Madison, 1987.

26. **Oesterhelt, D. and Stoeckenius, W.,** Rhodopsin-like Protein from the Purple Membrane of *Halobacterium halobium,* Nature New Biol. 233, 149, 1971.

27. **Albertsson, P.-A.** *Partition of Cell Particles and Macromolecules,* Almqvist & Wiksell (Stockholm), 1971.

28. **Ogrodnik, A., Volk, M., Letterer, R., Feick, R., and Michel-Beyerle, M. E.,** Determination of free energies in reaction centers of *Rb. sphaeroides, Biochim. Biophys. Acta,* 936, 361, 1988.

29. **Vanderkooi, G., Senior, A. E., Capaldi, R. A., and Hayashi, H.,** Biological Membrane Structure— III. The lattice structure of membranous cytochrome oxidase, *Biochim, Biophys, Acta,* 274, 38, 1972.

30. **Helenius, A. and Simons, K.,** Solubilization of Membranes by Detergents, *Biochim. Biophys. Acta,* 415, 29, 1974.
31. **Helenius, A., McCaslin, D. R., Fries, E., and Tanford, C.,** Properties of Detergents, *Meth. Enzymol.,* 56, 734, 1979.
32. **Welte, W., Leonhard, M., Diederichs, K., Weltzien, H.-U., Restall, C., Hall, C., and Chapman, D.,** Stabilization of detergent-solubilized Ca^{2+}-ATPase by poly(ethyleneglycol), *Biochim. Biophys. Acta,* 984, 193, 1989.
33. **Ford, R. C., Picot, D., and Garavito, R. M.,** Crystallization of the photosystem I reaction centre, *EMBO J.,* 6, 1581, 1987.
34. **Chang, H. W. and Bock, E.,** Pitfalls in the use of commercial nonionic detergents for the solubilization of integral membrane proteins: sulfhydryl oxidizing contaminants and thier elimination, *Anal. Biochem.* 104, 112, 1980.
35. **Hjelmeland, L. M., Nebert, D. W., and Osborne, J. C.,** Sulfobetaine derivatives of bile acis: nondenaturing surfactants for membrane biochemistry, *Anal. Biochem.* 130, 72, 1983.
36. **Hjelmeland, L. M. and Chrambach, A.,** Solubilization of functional membrane proteins, *Methods Enzymol.,* 104, 305, 1984.
37. **Timmins, P. A., Leonhard, M., Weltzien, H. U., Wacker, T., and Welte, W.,** A physical characterization of some detergents of potential use for membrane protein crystallization, *FEBS Lett.,* 238, 361, 1988.
38. **Caron, M. G., Srinivasan, Y., Pitha, J., Kociolek, K., and Lefkowitz, R. J.,** Affinity Chromatography of the β-Adrenergic Receptor, *J. Biol. Chem.,* 254, 2923, 1979.
39. **Michel, H.,** Crystallization of membrane proteins, Trends Biochem. Sci., *Sciences,* 8, 56, 1983.
40. **Henderson, R. and Shotton, D.,** Crystallization of Purple Membrane in Three Dimensions, *J. Mol. Biol.,* 139, 99, 1980.
41. **Hovmöller, S., Slaughter, M., Berriman, J., Karlsson, B., Weiss, H., and Leonard, K.,** Structural studies of cytochrome reductase - improved membrane crystals of the enzyme complex and crystallization of a subcomplex, *J. Mol. Biol.,* 165, 401, 1983.
42. **Schertler, G.,** Kristallisation von Bakteriorhodopsin. Charakterisierung des M-Intermediats im Kristall, Thesis, University of Munich, 1988.
43. **Garavito, R. M., Markovic-Housley, Z., and Jenkins, J. A.,** The Growth and Characterization of Membrane Protein Crystals, *J. Crystal Growth,* 76, 701, 1986.
44. **Michel, H. and Oesterhelt, D.,** Three-dimensional crystals of membrane proteins: bacteriorhodopsin, *Proc. Natl. Acad. Sci., USA,* 77, 1283, 1980.
45. **Garavito, R. M. and Rosenbusch, J. P.,** Three-dimensional crystals of an integral membrane protein: initial X-ray analysis, *J. Cell Biol.,* 86, 327, 1980.
46. **Michel, H.,** Three-dimensional crystals of a membrane protein complex. The photosynthetic reaction centre from *Rhodopseudomonas viridis, J. Mol. Biol.,* 158, 567, 1982.
47. **Allen, J. P. and Feher, G.,** Crystallization of reaction center from *Rhodopseudomonas sphaeroides:* preliminary characterization, *Proc. Natl. Acad. Sci. USA,* 81, 4795, 1984,
48. **Change, C.-H., Schiffer, M., Tiede, D., Smith, U., and Norris, J.,** Characterization of bacterial photosynthetic reaction center crystals from *Rhodopseudomonas sphaeroides* R-26 by X-ray diffraction, *J. Mol. Biol.,* 186, 201, 1985.
49. **Garavito, R. M., Hinz, U., and Neuhaus, J. -M.,** The Crystallization of Outer Membrane Proteins from *Escherichia coli, J. Biol. Chem.,* 259, 4254, 1984.
50. **Picot, D. and Garavito, R. M.,** Crystallization of a membrane protein: prostaglandin H synthase, in *Cytochrome P-450: Biochemistry and Biophysics,* Schuster, I., Ed., Taylor & Francis, London, 1989.
51. **Nestel, U., Wacker, T., Woitzik, D., Weckesser, J., Kreutz, W., and Welte, W.,** Crystallization and preliminary X-ray analysis of porin from *Rhodobacter capsulatus, FEBS Lett.,* 242, 405, 1989.
52. **Witt, I., Witt, H. T., Fiore, D. Di, Rögner, M., Hinrichs, W., Saenger, W., Granzin, J., Betzel, C., and Dauter, Z.,** X-ray characterization of single crystals of the reaction center I of water splitting photosynthesis, *Ber. Bunsenges. Phys. Chem.,* 92, 1503, 1988.
53. **Hawthornthwaite, A. M., Cogdell. R. J., Woolley, K. J., Wightman, P. A., Ferguson, L. A., and Lindsay, J. G.,** Crystallization and characterization of two crystal forms of the B800-850 light harvesting complex from *Rhodopseudomonas acidophila* strain 10050, *J. Mol. Biol.,* 209, 833, 1989.

Chapter 4

CRYSTALLIZING MEMBRANE PROTEINS: EXPERIMENTS ON DIFFERENT SYSTEMS

R. Michael Garavito

TABLE OF CONTENTS

I. INTRODUCTION

Integral membrane proteins possess extensive hydrophobic surfaces which allow them to function within or attached to a lipid bilayer. Their amphipathic nature has made them difficult to characterize biochemically and to crystallize for high resolution X-ray analysis. With the introduction of new, well-defined detergents into membrane biochemistry, methods for crystallizing integral membrane proteins could be developed[1-3] and reports of membrane protein crystals are now appearing in the literature.[4-12] The rapid development of crystallization methods for membrane proteins is not due to the use of radically new techniques; it is due to the growing awareness of the role detergents play in solubilizing membrane proteins and how the detergents can affect the crystallization process. As the species that is crystallized is a protein-detergent complex (PDC), understanding how these complexes behave in solution and during crystallization is necessary in order to rationally manipulate the system. The consensus among those who have been successful in obtaining membrane protein crystals is that finding conditions for microcrystal growth is "easy", but finding the conditions for optimal crystal growth is still very difficult. Much of the problem still lies in creating a detergent environment which is conducive to the growth of large crystals. In this chapter I will discuss some aspects of detergent behavior which can affect crystallization and my group's experiences in crystallizing a variety of integral membrane proteins.

II. DETERGENT CHEMISTRY: A PRIMER

A basic understanding of detergent chemistry is necessary when attempting any crystallization experiments on membrane proteins. Some older reviews cover the general aspects of detergent behavior as they apply to membrane biochemistry.[13-15] Reviews and monographs covering surfactant physical chemistry give more information applicable to crystallization experiments, but they are not written from a biochemical point of view.[16-18] Reports relating detergent physical chemistry to events observed during membrane protein crystallization are now appearing[19,20] and provide a good basis for understanding detergent behavior (see Chapter 2). Much of this research is reviewed by other authors in this volume. In this section, I would like to review briefly some aspects of detergent behavior that has influenced my groups' approach to crystallizing membrane proteins.

Detergents or surfactants in general are amphipathic molecules with distinct hydrophilic (head group) and hydrophobic (tail) moieties. The detergents used to crystallize membrane proteins (Table 1) are generally simple nonionic or zwitterionic surfactants: they have only a single polar head group and a single unbranched, saturated alkyl tail. At dilute concentrations, surfactants are readily soluble in water as a monomeric species, but at a critical concentration the monomers begin to self-associate into disordered, fluid aggregates called micelles (Figure 1A). The driving force for micellarization is the hydrophobic effect: there is a favorable free-energy change upon reducing the contact of the hydrophobic surfactant tail with water while maintaining the hydration of the headgroups. The critical micelle concentration (CMC) varies with the size and character of the surfactant tail and head group moieties.[16,17] For many detergents, other phase boundaries exist. Some detergents like alkyl *N*-methyl glucamides display a crystallization phase boundary. Many detergents exhibit a micellar phase separation behavior (see Chapter 2) where there is a partitioning of the detergent into detergent-rich and detergent-poor phases (Figure 2). As this phase transition occurs entirely within the region where micelles exist, most researchers consider it a micelle-dependent phenomenon where micelles can aggregate into large clusters (see Chapter 2) and/or fuse into larger micellar structures.[17,18,21] What truly does occur depends on the size and structure of the detergent. The relevance of this phase transition behavior to membrane protein crystallization is that the phasing "out" of detergents from an aqueous solution is

TABLE 1
Detergents Used in Membrane Protein Crystallization

Detergent[a]	Mol wt	CMC(mM)[b]	PEG[c]	AS[c]
β-Octyl glucoside (β-OG)	292	22—23	UC	LC
β-Nonyl glucoside	306	6.5	UC	LC
Decyl maltoside (C$_{10}$-M)	482	1.6	UC	LC
Dodecyl maltoside (C$_{12}$-M)	510	0.16	UC	LC
Octyl tetraoxyethylene (C$_8$E$_4$)	306	7.0	LC	LC
Octyl pentaoxyethylene (C$_8$E$_5$)	350	4.3	LC	LC
Dodecyl octaoxyethylene (C$_{12}$E$_8$)	518	0.08	LC	LC
Decyl dimethylamine-N-oxide (DDAO)	201	10.4	UC	?[d]
Dodecyl dimethylamine-N-oxide (LDAO)	229	1.1	UC	?
Dodecyl dimethylaminopropyl sulfoxide (Zwittergent 3-12)	335	3.0	UC	?
Octyl 2-hydroxyethylsulfoxide	206	29.9	UC	?
Nonanoyl N-methyl glucamide (MEGA-9)[e]	363	23.0	?	?
Decanoyl N-methyl glucamide (MEGA-10)[e]	377	5.0	?	?

[a] The generic name and one of the common abbreviations are given for each detergent. No consistent nomenclature is adhered to in the chemical and biochemical literature.

[b] The CMC will vary depending on the temperature, solution conditions and detergent purity. The CMCs reported here are from the literature or measured by my group.

[c] These columns represent the type of detergent phase separation boundaries observed in the presence of poly ethylene glycol (PEG) or ammonium sulfate (AS). UC refers to an upper consolute boundary while LC refers to a lower consolute boundary. See Figure 2 and References 18 and 25 for more explanation.

[d] The question marks in the PEG and AS columns indicate that no detergent phase separation has been observed, though no extensive study has been made.

[e] The MEGA-9 and MEGA-10 detergents will crystallize out of solution below 10 and 18°C, respectively, depending on the solution conditions. They also tend to easily supersaturate which can compound the problems with their use.

strongly affected, or even induced by the presence of salt or polymer solutes.[19,20] Hence, the compounds used for crystallization not only can ''salt-out'' the protein[22] but also the detergent. This has been a major factor affecting our design of crystallization experiments.

At higher detergent concentrations (above 25% w/v) micellar detergent phases or mesophases appear.[20,23] Under most conditions, a membrane protein preparation would never be exposed to such high detergent concentrations, even during most crystallization experiments. However, as crystal formation requires the close approach of PDCs to create ordered aggregates, the local concentration of detergent in pre-crystal nuclei and in the final crystal can be quite high. For OmpF porin crystals, the estimated β-octyl glucoside concentration in the crystals varies from 30 to 40% (w/v), depending on the crystal form.[5] If one considers only the crystal volume not occupied by the protein, the detergent concentration then increases to about 40 to 60%. Thus the kinds of mesophases and mesophase transitions exhibited by a detergent must be important in crystal nucleation and growth.

Micellar structure and stability, mesophase existence and mesophase stability depend on the structure and character of the surfactant. Simple qualitative relationships based on the effective surface area of the surfactant can be used to predict micelle shape and potential mesophase formation.[23-25] The kind of aggregate shapes which are energetically feasible for packed surfactants would depend on the balance between hydrocarbon chain/water repulsion and head group repulsion (a combination of steric, electrostatic and solvation effects). For the nonionic and zwitterionic detergents in Table 1, the micelle shapes are either small spheres or oblate discs. Using packing considerations for individual surfactants,[24,25] an average number of monomers per micelle (the aggregation number) can be determined. The calculated values can then be compared with those derived from the average micellar size as determined by a number of techniques including molecular sieve chromatography.[37] Both

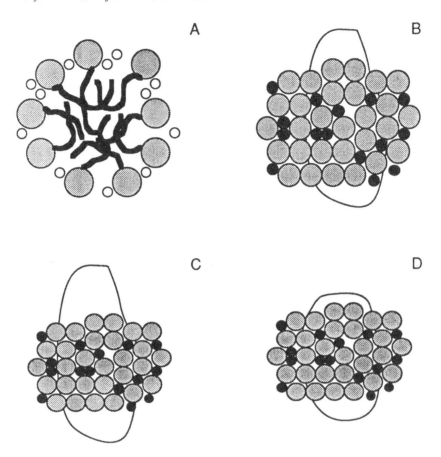

FIGURE 1.   Schematic diagrams of micelles and protein-detergent complexes (PDCs). In (A), detergent monomers aggregate to form a micelle where their fluid hydrophobic tails pack in a disordered manner to minimize their contact with water. Their polar head groups (shaded circles) still interact with bulk solvent and allow some water (open circles) to penetrate partially into the micelle. A protein-detergent complex, shown in (B), would have a layer of detergent bound to the hydrophobic surface of the protein so that the detergent's polar headgroups (lightly-shaded circles) face towards the bulk solvent. It is supposed that the hydrophilic surface of the protein would be exposed to the aqueous solvent, though portions may be concealed due to the size of the detergent layer. The darkly shaded circles represent other molecules (water, lipids, organic solvents) which are also solubilized in this complex. There is no apparent order to the detergent layer and it is still fluid. In (C) and (D), making the detergent layer smaller, by perhaps altering the detergent, could make more protein surface available for crystallization, as between (B) and (C), or expose enough protein surface of a small membrane protein (D) to allow protein-protein contacts. In the latter case, we would expect significant detergent effects in the crystallization process.

calculated and experimental aggregation numbers for small detergents (like those listed in Table 1) fall between 40 to 100 molecules per micelle. For mesophases, the same structural (packing) arguments can be made: the effective shape of a surfactant will determine how energetically feasible it can be to pack the monomers into the requisite shapes and how easy it will be to transform one phase into another.

For both micelles and mesophases, packing forces can create surfaces within which the surfactant monomers may not pack well locally and disrupt some polar head group or hydrophobic tail interactions. There are also aspects of detergent monomer dynamics and fluidity where local fluctuations could create transient changes in radii of curvature in micelles or mesophases and affect surfactant monomer exchange with the solution.[18] Despite the

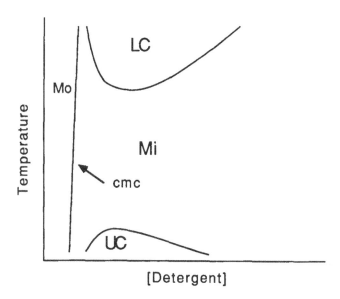

FIGURE 2. A schematic of a temperature vs. detergent concentration phase diagram showing the behavior of an alkyl glucoside detergent (like β-OG) mixed with an alkyl oligo-oxyethylene detergent (like C_8E_5). The CMC boundary shows the transition from detergent monomers (Mo) to micelles (Mi). PEG induces alkyl glucosides to form an upper consolute phase boundary (UC) where cooling the system causes a phase transition into two liquid phases (as in crystallization experiments with OmpF porin). The addition of the alkyl oligo-oxyethylene detergent creates a low consolute boundary (LC) at higher temperatures, but also suppresses the UC boundary. Thus, by adjusting the ratio between the two detergents, the two phase boundaries can be manipulated to create a phase-free region for crystallization.

fluctuating, fluid nature of these detergent aggregates, micelles are relatively discrete entities. In dilute solution, micelles of small detergents do not fuse or change size with changes in temperature or detergent concentration until a phase boundary is reached.[25] Hence micelles can be considered a distinct colloidal species which have their own unique properties.

As depicted schematically in Figure 1A, the head groups in the micelle do not completely shield the hydrophobic core from the bulk solvent: water has significant access to the surface of the hydrophobic core.[25] This has consequences for solubilizing amphipathic co-solutes like small amphiphiles. Micelles of small nonionic or zwitterionic detergents (Table 1) readily solubilize polar organic molecules like short chain alcohols.[26,27,28] Substantial co-micellization is observed as evidenced by macroscopic and microscopic changes in micelle behavior. The addition of either ethanol or dioxane to aqueous solutions of the dodecyl polyoxyethylene detergent $C_{12}E_{23}$ initially causes a decrease in the apparent CMC due to energetically favorable solubilization.[28] As polar organic co-solute is added to high concentrations, the apparent CMC then begins to increase and the measured micelle aggregation number decreases, indicating the breakdown of the micelle structure. These mixed micelles may exhibit substantial differences in apparent shape, size, phase formation, and phase stability[16,27,28] compared to pure micelles. However, no simple method exists to predict how a given co-solute will effect the structure and behavior of a given surfactant micelle. Hence, the use of co-solutes in membrane protein crystallization (see Section III) cannot be explained by simple effects on micelle size or shape.[4,6]

III. THE PROTEIN-DETERGENT COMPLEX

The species we want to crystallize is a protein-detergent complex (PDC) which is soluble in aqueous solution by virtue of the bound detergent (Figure 1B). This simplistic picture does not convey the difficulties one confronts in producing a homogeneous, monodisperse preparation of a membrane protein. After disrupting native membranes by the addition of detergent, a vast array of protein-lipid-detergent aggregates exists. Purifying this heterogeneous system to create PDCs of defined size and character means that unwanted protein and lipid contaminants must be removed without inactivating the protein of interest. Our experiences with OmpF porin[5] and other membrane proteins[6,10] clearly show that protein or lipid contaminants can adversely affect the crystallization of these proteins. Purity, as measured by the homogeneity and monodispersity of the PDC, is critical for the crystallization of membrane proteins. If successful crystallization has eluded a researcher, he or she should consider the purification scheme as the primary source of the problem.

When a sufficiently homogeneous preparation is obtained, one should now begin to consider the nature of the PDC and how changing the nature of the detergent can affect its characteristics. A few studies have experimentally determined the distribution of detergent about a solubilized membrane protein (M. Zulauf and J. P. Rosenbusch, personal communication). The detergent monomers apparently bind to the proteins hydrophobic surface to create a torus-like layer about it (Figure 1B). Though it has not been experimentally proven, it is assumed that surface covered by the detergent is primarily the membrane-embedded portion of the protein. The likely mode of interaction between the protein and detergent is a process analogous to micellarization where the proteins transmembrane surface acts as a nucleus for micelle-like detergent association.[17,29] However, the hydrophobic surface on proteins may prevent efficient, low free-energy detergent binding: the transmembrane region of most membrane proteins are designed to interact with lipid in a bilayer state. Furthermore, the surface available for detergent binding would not necessarily be uniform, structurally or compositionally. Hence, spatial constraints on detergent binding can arise due to the shape and extent of the transmembrane surface and detergent monomers may, therefore, be forced to adopt sterically unfavorable binding modes on a PDC that would not occur in pure detergent micelles. This may create patches of unstably-bound detergent or actual discontinuities in the detergent layer. Metastable or unstable PDCs are often observed in membrane biochemistry and have indicated, in the past "poor" solubilization by a specific detergent. If regions of the detergent surface are unstable, as can be the case at limiting detergent concentrations or for poorly solubilized proteins, nonspecific protein interactions can produce slow protein aggregation leading to precipitation or, perhaps, protein micelle formation.[15] This phenomena may be a major source of polydispersity in membrane protein preparations. Protein samples used for crystallization should be free of such aggregation.

Stable protein-detergent aggregates can experience other types of intermolecular interactions which affect the behavior of PDCs in solution. Strong protein-protein interactions are very common and can be readily induced by the addition of salt (NaCl or ammonium sulfate) or polar polymers like polyethylene glycol. This solute-induced aggregation has been used in the purification of membrane proteins[30,31] It further indicates that the classical methods for inducing protein crystallization[22] (see Chapter 1) are effective for membrane proteins. The difference between membrane and soluble proteins is that the detergent layer of the PDC can also induce various types of interparticle interactions. As mentioned above pure detergent micelles can be "salted out" of solution by the addition of salt and other solutes or by increasing temperature.[19-21] The resulting phase change produces a viscous, detergent-rich micellar phase which co-exists with an aqueous detergent-poor phase. Identical detergent-induced phase behavior is exhibited by PDCs at high salt or polymer concentrations.[5,19,20] The Triton X-114 separation technique described by Bordier[32] is an example of how PDCs are particularly affected by detergent phase phenomena. Hence, we can see that a PDC remains soluble due to repulsive forces arising from *both* the protein and detergent head group surfaces. To crystallize such PDCs, we must reduce the global interparticle repulsion to a point where protein-protein interactions begin to occur as part of crystal nucleation.

We have discussed the various aspects of the structure and behavior of PDCs necessary for their crystallization. Unfortunately, protein preparations ideal for membrane protein crystallization cannot always be obtained. Often membrane protein preparations contain variable amounts of lipid contaminants. As lipids often bind tighter to membrane proteins than do detergents and native lipids are very heterogeneous, lipid contamination can thus create a very heterogeneous population of PDCs. Adding to the problem, many detergents unsuitable for general membrane protein crystallization (e.g., Triton X-100, Tween-20, or Brij 35) are more effective in disrupting native membranes and in solubilizing membrane proteins than detergents useful for crystallization. Large scale purification schemes often use "less desirable", but generally cheaper detergents for many of the initial purification steps. The first detergent must then be exchanged for one which is more suitable for crystallization. Again, the potential for detergent contamination in the final preparation is high: the heterogeneous nature of these "purification" detergents and their low CMCs can result in a very heterogenous PDC population even at low levels of contamination. The big lesson to be learned is that a great deal of time must be invested in preparative biochemistry in order to produce material suitable for crystallization. Because the lipid and detergent contaminants are always interacting with the PDCs, changing their size, structure, and behavior, the determination of acceptable levels of contamination for crystallization is an important endeavor. In the next section, we will discuss our practical experiences in crystallizing membrane protein that perhaps offer insights to others in how to approach a new, untried, membrane protein.

IV. CRYSTALLIZATION STRATEGIES AND PROTOCOLS

With the preceding information and concepts, we can develop a basic strategy for crystallizing membrane proteins. In our group, we generally address four aspects of the problem before we begin. First, we determine if the protein preparation is chemically pure and physically homogeneous. Heterogeneous and polydisperse preparations of proteins do not crystallize well even for soluble proteins.[33] Second, we determine if the protein is in an active or, at least, a defined conformational state. Many times, detergent solubilization can lead to reversible and/or irreversible inactivation of the protein. It is important to ensure that the protein preparation is conformationally homogeneous, regardless of whether the protein is active or not. We were able to crystallize a low pH conformational variant of OmpF porin[34], but only after defining the range of conditions where it existed. We have followed similar experimental designs to access protein stability and activity. De Grip has provided an excellent account of surveying the stability of visual rhodopsin in the presence of a variety of detergents.[43] Third, we monitor the solubility of the PDC at high protein concentration in the presence of low concentrations of crystallization agents. While many membrane proteins are soluble at low concentrations in detergent (below 1 mg/ml), higher concentrations sometimes induce nonspecific aggregation that is exacerbated by ammonium sulfate or polyethylene glycol (PEG). This phenomenon is distinguished from the desired "salting-out" effect by the slow and progressive nature of the precipitation. In such cases, we have had to substantially alter the solution conditions (i.e., pH, ionic strength, etc.) or the solubilizing detergent before a stable precrystallization preparation could be achieved. Lastly, the desired range of precipitant concentrations is tested for its compatibility with a particular detergent to avoid unwanted detergent phase transitions. A poor choice of conditions can lead to the crystallization of the detergent or extreme detergent phase separation throughout most of the experiments. Testing detergent/precipitant combinations beforehand will help minimize the problem. (There are differences in the detergent-dependent behavior between pure detergent micelles and PDCs but these cannot be predicted beforehand.)

We have addressed the question of how to choose suitable detergents for membrane crystallization but mainly from the viewpoint of obtaining pure, homogeneous (monodisperse), and active protein preparations. However, choosing a detergent specifically for crystallization experiments is still quite an empirical endeavor. My group has succeeded in obtaining crystalline preparations of several membrane proteins using the detergents listed in Table 1. Our basic criteria were that the detergents should (1) be pure, (2) produce small, monodisperse micelles, and (3) be chemically stable and inert. Many nonionic and zwitterionic detergents satisfy these criteria. The behavior of their micellar solutions should also be relatively constant to over a wide range of pH and ionic strengths. This is an important point to consider when designing crystallization experiments.

The underlying assumption in our selection of detergents is that by using detergents which would form small micelles we would create the smallest PDC possible (Figure 1B and 1C). The goal is to maximize the exposed protein surface for making crystal contacts without sacrificing protein stability or solubility. Hence, small, discrete PDCs would then be able to approach each other to form ordered aggregates and crystal nuclei with minimal interference from the detergent surface. One can extrapolate from the simple caricatures of the situations shown in Figure 1 to two extremes. If the protein is quite embedded in the membrane, it might have very little protein surface exposed in a PDC and, therefore, detergent phenomena would dominate the crystallization process. We feel that this is the case for OmpF porin crystallization. On the other hand, if there is an extensive extramembranous protein surface, as in a large protein complex, the detergent region accounts for only a minor part of the total surface and may have a minimal influence on the crystallization process in a significant way. For truly large protein complexes, one might expect that crystals could be obtained using large, heterogeneous detergents like Tween-20 or Triton X-100. In two recent cases this seems to be the case.[10,12]

For the rest of this section, we will describe the crystallization experiments and results for five different membrane proteins. In all cases small well-formed crystals were obtained after about 1 month of effort, though refining the conditions to obtain large crystals took much longer. The discussions should be used only as a guide for approaching a new protein. Our experience has been that once the detergent problems have been handled adequately, the general methods for screening crystallization conditions work quite well (see Chapter 1).[22]

A. OMPF PORIN

We have explored a great variety of conditions in our experiments on OmpF porin and have defined conditions where we can obtain several different crystal forms (Figure 3). Most of this work has been reviewed elsewhere[5] and I will summarize only the major points. The transmembrane channel-forming protein OmpF porin is a very hydrophobic, membrane-embedded protein[35] which has a surprisingly polar amino acid sequence. This protein does not precipitate in the presence of high salt (ammonium sulfate or phosphate) or polyethylene glycol (PEG) concentrations. Generally, detergent phase separation occurs first and the protein partitions with the detergent-rich phase. As this process concentrates the protein (to about 200 mg/ml), the detergent-rich layer becomes extremely viscous. Over a wide pH range, crystals eventually grew in the detergent-rich phase in the presence of ammonium sulfate (~70% sat.) or PEG 4000 18 to 25%). We tested α- and β-octyl glucoside and found that both allowed porin crystallization. However, as α-octyl glucoside (α-OG) crystallizes at an extremely high temperature ~42°C), the β anomer was chosen as the best detergent.

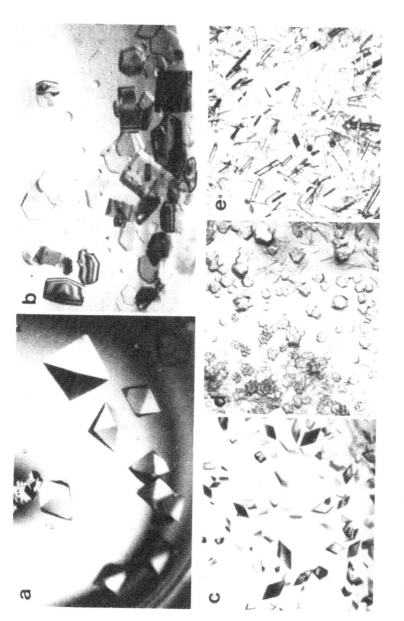

FIGURE 3. OmpF porin crystals. Panels a and b show the tetragonal and hexagonal crystals grown in β-OG/PEG/NaCl systems. Panel c shows the triclinic crystals grown in the presence of octyl 2-hydroxyethylsulfoxide. Panels d and e show the effect of micelle perturbants on crystallization. Conditions which yield a mix of hexagonal and monoclinic crystals can be induced to form only monoclinic crystals in the presences of 1% (v/v) dioxane. (From Garavito, R. M., Markovic-Housley, Z., and Jenkins, J. A., *J. Crystal Growth*, 76, 701, 1986. With permission.)

Eventually most of the detergents in Table 1 were tested and crystals obtained but β-OG always gave the best results.[2]

The design of our initial vapor diffusion experiments almost always resulted in, first, phase separation and then crystallization.[5] Thus, we had concluded that the phase separation phenomenon was obligatory for crystallization. However, upon our development of a microdialysis system for porin crystallization, it became clear that crystals could be obtained without phase separation but only when the conditions were relatively close to the phase separation boundary.[2] The microdialysis system allowed us to explore, in a systematic manner, the influence of each component on the system. We observed several distinct solute effects.[2] First, moderate changes on the ionic strength were the primary determinant in selecting which of three crystal forms would grow. Second, changes in detergent concentration affected only the speed of nucleation and crystal growth, not crystal habit or quality. The slowest, most uniform crystal growth conditions occurred when the detergent concentration was only about 30% higher than the CMC. Third, the sensitivity of crystal growth and integrity to changes in temperature depended on the detergent used. For β-octyl glucoside, crystal nucleation and growth speed increased markedly with decreasing temperature. This effect can be reduced and even reversed by using noncarbohydrate based detergents (e.g., C_8E_5). The details of these effects strongly suggest that detergent-detergent interactions play a critical role in crystal nucleation and growth. This led us to use controlled temperature changes and detergent mixtures to fine-tune the crystallization process. Lastly, certain detergents and detergent mixtures could override the selection of the three primary crystal forms by solution ionic strength. In Figure 3c, triclinic crystals appeared when β-octyl glucoside was replaced with octyl 2-hydroxyethylsulfoxide. If detergents with dodecyl alkyl tails were used, the tetragonal to hexagonal transition could be induced by doping dodecyldemethylamine oxide (LDAO) with $C_{12}E_6$. These experiments suggested to us that the nature of the detergent head group could profoundly affect how the PDC could pack in a crystal. Hence, the detergent surface on a PDC participates, in some manner, in the crystallization process.

The system was finally optimized to produce large (0.7 × 0.7 × 0.5 mm) tetragonal crystals which we felt would yield the best diffraction data for analysis[2] (Figure 3a). Using the microdialysis technique, we kept the ionic strength high throughout the experiment while limiting the total detergent concentration. PEG 2000 was used instead of PEG 4000 which resulted in a less abrupt crystallization threshold and an increased distance between the porin crystallization and detergent phase separation boundaries. The chances for detergent phase separation was further reduced by adding a second detergent C_8E_x (where x = 6 to 11) which suppressed the β-OG induced phase behavior (see Figure 2).[2] After crystal growth, the microdialysis technique also allowed us to equilibrate the crystals against new buffers for experimentation or long-term storage. At all times, the crystals were very sensitive to temperature changes and alteration in their detergent environment.

Though the tetragonal crystal form was most desired for our X-ray crystallographic work, we also refined the conditions for obtaining the hexagonal and monoclinic forms. In these cases, we were less successful for two reasons. First, the crystal nucleation and growth rates could not be as well controlled as in the experiments on the tetragonal form. Thus, the size and number of crystals were always extremely variable from experiment to experiment. Second, the crystallization boundary between the two crystal forms was less distinct than that between the tetragonal form and the others. Hence, the two forms were often intermixed and could poison the growth of each other. Nonetheless, we found that the conditions for monoclinic crystal growth could be refined using a range or organic co-solutes with micelle perturbing characteristics.[3] Michel calls similar compounds (like 1,2,3-heptane triol) ''small amphiphiles'' (see Chapter 3).[1]

Additions of ethanol, dioxane, or 1,6-hexanediol can alter the crystallization process at conditions where hexagonal and monoclinic crystals grow so that only the monoclinic form is seen (Figure 3).[3] The effective concentrations of the micelle perturbants are between 1 to 5% (v/v). Higher concentrations of the compounds invariably led to more rapid crystallization, excessive microcrystal formation and, finally, irreversible protein precipitation. These effects at high concentrations seem to be due to the lowering of the detergents capacity to solubilize protein (i.e., its apparent CMC rises) which then renders the PDC more prone to aggregation. We have noted that the less polar compounds (alcohols) act at lower concentrations than the more polar compounds (like diols or triols). Michel's very polar small amphiphile 1,2,3-heptane triol[1] had little effect in our systems. However, it must be remarked that Michel's system for selecting efficacious compounds relied on crystallization experiments using ammonium sulfate. We found that many micelle perturbants which were efficacious in PEG systems were insoluble in ammonium sulfate. Thus the compatibility of the micelle perturbants with the precipitation/crystallization agent may be a factor in their successful use.

In summary, the crystallization behavior of OmpF porin displays characteristics that are suggestive of a dominant role of the detergent in crystal nucleation and growth. This does not mean to imply that the detergent is responsible for crystal order, but rather that positive and negative detergent effects strongly influence the kinds of protein-protein interactions which can occur during crystallization. As porin is relatively embedded in the membrane,[35] its crystallization behavior may be typical of smaller or more buried membrane proteins.

B. PHOSPHOPORIN (PHOE)

PhoE is another of the porin class of outer membrane passive transport channel. It is evolutionarily related to OmpF porin, though it has developed a selectivity for inorganic and organic phosphates.[35] Like many porins, its native form in the membrane and in the solubilized state is a trimer. Using protein kindly supplied to me by J. P. Rosenbusch, I began a search for crystallization conditions (using microdialysis), around those developed for OmpF porin. With a basic PEG 4000 and β-OG system (1.0% β-OG, 20 mM sodium phosphate, pH 7.0) a variety of crystals were obtained as the experiments were supplemented with NaCl. The crystal grew throughout a broad range of PEG concentrations (12 to 14%). At low NaCl concentrations (less than 50 mM) large, thin hexagonal plates grew. These crystals were well-ordered and diffracted X-rays out to at least 3Å. The tentative space group is P6$_3$22 and the lattice constants are similar to those of the equivalent hexagonal space group found for OmpF porin.[5] At higher NaCl concentrations, orthorhombic prisoms grew as well as small lozenge-shaped microcrystals of undefined character. Thus, this sister protein to OmpF porin behaves similarly (but, by no means, identically) under the general crystallization conditions developed for OmpF porin. These crystals are currently being improved by others (A. Tucker, EMBL, Heidelberg; N. Koenig and R. Pauptit, Biocenter, Basel).

C. MALTOPORIN (LAMB)

The lambda phage receptor in *E. coli* is a passive diffusion pore with a pronounced selectivity for maltose and maltodextrins.[35] There is no apparent evolutionary relationship between LamB and the other porins, though LamB does exist in the membrane as a trimer. We felt that LamB would behave in a similar manner as OmpF porin: the membrane embedded nature of the protein should result in significant detergent effects during crystallization. This was borne out in our first crystallization trials using vapor diffusion (Figure 4).[6] For technical reasons, the initial trials were done with a C$_8$-POE/β-OG detergent mixture. (C$_8$-POE has a variable oxyethylene headgroup length and can be considered a polydisperse version of C$_8$E$_4$ or C$_8$E$_5$ in Table 1). With this detergent system, crystals grew at relatively high PEG 4000 and NaCl concentrations (25% and 0.7 M, respectively). Detergent phase separation occurred well before bipyramidal crystals appeared. These crystals did diffract and were

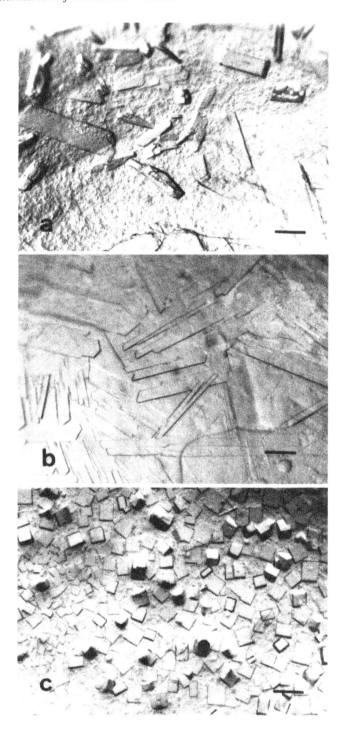

FIGURE 4.   Crystals of LamB (maltoporin). Chunky prisms can be grown in C_8E_4/PEG/NaCl system (a). By adding a small amount of β-OG to the crystallization medium, the general quality of the crystals improve (b). If instead a micelle perturbant was added (in this case, ethylene glycol butyl ether) a new crystal form grew. (From Garavito, R. M., Hinz, V., and Neuhaus, J. M., *J. Biol. Chem.*, 259, 42, 1984. With permission.)

generally suitable for X-ray diffraction analysis.[6] Crystals could be obtained at much lower PEG and salt concentrations if a micelle perturbant was added, though the space group changed.[6] It was then decided to try the detergent C_8E_4 as it would afford more homogeneous PDCs. As shown in Figure 3, a pure C_8E_4 system allowed the growth of chunky, prismatic crystals in the presence of PEG 4000. Shifting to a C_8E_4/β-OG mixed system brought some improvement in crystal shape and quality (Figure 4b). Good X-ray diffraction was observed despite the relative thinness of the crystals (0.04 to 0.06 mm). Again, if micelle perturbants were added, a new crystal form grew.

The LamB crystallization trials demonstrated that crystals could be obtained under a number of conditions. However, detergent effects like phase separation were observed throughout the successful trials. The next step for improving the crystals would be to shift to a microdialysis system to control more accurately the detergent environment. Micelle perturbants were again successful in inducing the growth of new crystal forms, though the more polar compounds like 1,2,3-heptane triol were least effective. Experiments to improve the crystal quality are underway elsewhere (A. Tucker, EMBL, Heidelberg; K. Stauffer and R. Pauptit, Biocenter, Basel).

D. PROSTAGLANDIN H SYNTHASE (PGS)

The conversion of arachidonic acid to prostaglandin H is catalyzed by the membrane bound enzyme prostaglandin H synthase.[38] This glycosylated enzyme uses a noncovalently bound heme as a cofactor in both the cyclooxygenase and hydroperoxidase reactions. PGS can be easily isolated from a microsomal preparation as a stable dimer ($2 \times 70,000$ Da) using nonionic detergents.[39,40] We have merged and modified the published purification schemes[39,40] so as to produce large amounts of pure, high active enzyme for crystallization experiments.[44] Unlike the previous membrane proteins we had studied, PGS has a relatively large extramembranous domain[41] Our experience with handling the protein during purification suggested that detergent effects would not play a great role in crystallization if a suitable detergent was found. For example, PGS could be easily and cleanly precipitated from detergent-containing solutions with PEG or ammonium sulfate without detergent phase separation (R. M. Garavito, unpublished observations), suggesting that the PDCs can be induced to aggregate well away from any detergent phase boundaries. However, two other detergent-related problems arose. First, the original purification scheme relied on Tween-20 to keep the enzyme solubilized and, therefore, used a final detergent exchange step to replace Tween-20 with a more desirable detergent. We found that the final molecular sieve column could not remove all of the unwanted detergent due to its very low CMC, but also that it could not even reduce the Tween-20 contamination to a reproducibly low level. Extreme variations could be seen from preparation to preparation. Part of this variation might have been due to a variable lipid content of the material obtained from the preceding purification steps. Nevertheless, the initial preparations contained a significant level of Tween-20 and lipid contamination. The second problem concerned a slow aggregation behavior of a certain fraction of the protein which led to its irreversible precipitation. As more Tween-20 was removed from the preparation, we found that this precipitation became more noticeable particularly when preparing protein for crystallization experiments with PEG. Part of the protein precipitated, even at low PEG 4000 concentrations, while the rest remained stably in solution. Was this behavior due to inadequate solubilization by small non-ionic detergents or is the protein still heterogeneous in some manner? We are currently examining these possibilities.

Despite these problems, PGS crystallizes readily in either PEG 4000 or ammonium sulfate (Figure 5). The best and most numerous crystals were obtained in the pH range 6.5 to 7.0 and at detergent concentrations near the CMC. There is a clear inverse correlation between the levels of lipid and Tween-20 contamination and the size and quality of the

FIGURE 5.   Crystals of prostaglandin synthase. In the upper panel, poor, needle-like crystals grew in the presence of heavy Tween-20 and lipid contamination. Note the gel-like protein-detergent-lipid liquid phase. In the middle panel, protein with lower Tween-20 and lipid contamination yields tetragonal rods, though the growth characteristics and crystal order is not optimal. In the lower panel, protein purified only in C_{10}-M and crystallized in β-OG gives only an orthorhombic form.

resulting crystals (Figure 5). At least five different crystal forms have been obtained and two have been characterized by X-ray diffraction. Crystals can be grown either by vapor diffusion or by microdialysis. Unlike the porins, the range of PEG 4000 concentration where crystals grow is quite broad (8 to 16%). No detergent phase separation is observed except at the highest PEG concentrations or in the presence of considerable Tween-20 and lipid contamination (upper panel, Figure 5).

One clear observation we made about PGS crystallization is that crystal quality always decreased when the ionic strength was too low. We observed similar phenomena during screening in experiments for all of the membrane proteins we have studied. When the crystallization medium was supplemented with 50 to 200 mM NaCl, better crystals grew than with no NaCl addition. Thus when attempting crystallization experiments with PEG, it may be advisable to start with a slightly elevated ionic strength.

The obvious solutions to the Tween-20 problem would be to use either a more effective detergent exchange procedure (e.g., affinity chromatography) or another more easily exchangeable detergent. Because of technical difficulties with the former solution, we explored the use of decyl maltoside (C_{10}-M) as the solubilization and purification detergent. C_{10}-M solubilizes PGS well and retains the enzyme's activity at all concentrations above the CMC. Furthermore, the CMC is high enough (see Table 1) to allow facile exchange with most of the other detergents, yet low enough to minimize the cost of its use. PGS purified in C_{10}-M had a minute lipid content and crystallized readily as large, orthorhombic prisms (lower panel in Figure 5; see Reference 44). With a preparation free of unwanted detergent we have examined the crystallization behavior of PGS in the presence of other detergents and detergent mixtures. PGS is much less sensitive to the detergent environment than OmpF porin, though detergent effects can be seen, primarily in nucleation rates, crystal size and crystal habit.

E. PHOTOSYSTEM I REACTION CENTER

As in plants, cyanobacteria have two photosystems in their photosynthetic apparatus: photosystem II which is involved in oxygen evolution and H^+ pumping and photosystem I which provides reducing equivalents for the production of NADPH.[42] Intact and active photosystem I reaction center (PS1-rc) complexes can be easily isolated from a number of cyanobacteria. From the thermophilic cyanobacterium *Phormidium laminosum*, we have isolated a 450 kDa complex which is stable in a number of detergents including SDS[10]. The PS1-rc is an extremely heterogeneous complex consisting of at least four distinct polypeptides, iron-suffer centers, and about 60 light-harvesting pigment molecules. The proper choice of detergent is very important here as we wish to keep this complex intact. Some small detergent like β-OG are inadequate for this task and result in pigment loss. Instead, the PS1-rc complex was generally handled in Triton X-100 or dodecyl maltoside (C_{12}-M).

In the presence of PEG 6000 and Mg^{++} ions, PS1-rc reversibly precipitates out of solution without loss of activity. By exploiting this behavior, we succeeded in preparing crystals of PS1-rc. The initial experiments were done using vapor diffusion, though larger and better quality crystals were obtained by free-interface diffusion[22] (see Chapter 1 for a description of this method) with crystallization being initiated by the layering of $MgCl_2$ onto a lower protein phase (8 mg/ml PS1-rc, 50mM Tris. HCl, 0.1% C_{12}-M and 8% PEG 6000). The major obstacles to crystallization seemed to be the rapid aggregation of PS1-rc to produce either numerous crystal nuclei or globular precipitation. Attempts to slow the nucleation process for better, more controlled crystal growth have not been entirely successful. Two different crystal forms have been observed: a large chunky prismatic form which displays birefringence and linear dichroism[10] and small well-shaped bipyramids. The largest crystals we have obtained do diffract X-rays to low resolution (8 Å) but their overall crystalline order has not been fully characterized (R. Ford and R. Pauptit, personal communication). The PS1-rc from a different species of cyanobacteria also shows a strong tendency to crystallize even at very low protein concentration.[11]

Though better PS1-rc crystals grow in the presence of C_{10}-M or C_{12}-M, the detergent Triton X-100 does not hinder crystallization. The large size of the PS1-rc complex seems to accommodate the more heterogeneous nature of the detergent and the relatively large detergent surface without preventing the protein-protein contacts necessary for crystallization.

Thus, if small detergents like β-OG tend to dissociate a protein complex, as in the case of PS1-rc, using a larger detergent may be a very reasonable alternative. Bovine heart cytochrome *c* oxidase is another large protein complex which is sensitive to small detergents.[37] Yoshikawa et al. have recently reported the growth of cytochrome *c* oxidase crystals from solutions containing very heterogeneous detergents like Tween-20 or Brij 35.[12] Therefore, the range of suitable detergents for crystallization can vary depending on the size and nature of the protein. I would not recommend the routine use of heterogeneous detergents for crystallizing large complexes. There are several better alternatives (e.g., $C_{12}E_8$, etc.) which would allow more control over the detergent environment.

V. SUMMARY

Our experiences with crystallization of membrane proteins has emphasized the need for having good information about the purity, stability, and behavior of a protein in detergent solutions. There exist already a good number of suitable detergents for crystallization experiments and several more promising detergents remain to be characterized. Thus, if stable, homogeneous protein preparations can be obtained, the chances of finding a suitable protein/detergent combination that can yield crystals is quite good. With the array of detergents and micelle perturbants currently available, there are a number of possible avenues for refining crystallization conditions. This also means that the number of significant variables in a membrane protein crystallization experiment is quite large.

We find that PEG/NaCl systems have been most successful in producing crystals of membrane proteins regardless of the detergent used. Other groups have also had good success with PEG,[7-9,11] but this does not necessarily imply that PEG is superior to ammonium sulfate for membrane protein crystallization as the number of successful cases is still too few. As scanning several protein/crystallization agent combinations is not extraordinarily time-consuming, we found a wide-range of conditions can be quickly explored. Thus there is no reason to ignore one crystallization agent or detergent in preference to another. For our work, the major investment of time and effort has been in protein preparation and we feel that this is where success in crystallization will be determined.

ACKNOWLEDGMENTS

I would like to thank those who have contributed to the thoughts and ideas expressed in this article. In particular, I am indebted to Drs. G. J. T. Tiddy and M. Zulauf for discussions on detergent physical chemistry. I would also like to acknowledge those who have contributed to the crystallization research in my laboratory through collaborations and collegial discussions, including A. Tucker, Drs. D. Picot, J. A. Jenkins, Z. Markovic-Housley, R. Ford, R. Pauptit, and J. P. Rosenbusch. I would also like to thank N. Campobasso for technical help in determining some detergent phase boundaries. Part of this work has been supported by the Swiss National Foundation (Grant 3.655.84 to R.M.G.), the Geigy-Jubiläum Stiftung, and the American Cancer Society (Grant BC-581).

REFERENCES

1. **Michel, H.**, *Trends Biochem. Sci.*, 8, 58, 1983.
2. **Garavito, R. M. and Rosenbusch, J. P.**, *Method Enzymol.*, 125, 309, 1986.
3. **Garavito, R. M., Markovic-Housley, Z., and Jenkins, J. A.**, *J. Crystal Growth*, 76, 701, 1986.
4. **Michel, H.**, *J. Molec. Biol.*, 158, 567, 1982.
5. **Garavito, R. M., Jenkins, J. A., Jansonius, J. N., Karlsson, R., and Rosenbusch, J. P.**, *J. Molec. Biol.*, 164, 313, 1983.
6. **Garavito, R. M., Hinz, U., and Neuhaus, J. M.**, *J. Biol. Chem.*, 259, 42, 1984.
7. **Allen, J. P. and Feher, G.**, *Proc. Natl. Acad. Sci. USA*, 81, 4795, 1984.
8. **Chang, C. H., Schiffer, M., Tiede, D., Smith, U., and Nororis, J.**, *J. Molec. Biol.*, 186, 201, 1985.
9. **Welte, W., Wacker, T., Leis, M., Kreutz, W., Shinozawa, J., Gad'on, N., and Drews, G.**, *FEBS Lett.*, 182, 260, 1985.
10. **Ford, R., Picot, D., and Garavito, R. M.**, *EMBO J.*, 6, 1581, 1987.
11. **Witt, I., Witt, H. T., Gerken, S., Saenger, W., Dekker, J. P., and Rogner, M.**, *FEBS Lett.*, 221, 260, 1987.
12. **Yoshikawa, S., Tera, T., Takahaski, Y. Tsukihara, T., and Caughey, W.**, *Proc. Natl. Acad. Sci. USA*, 85, 1354, 1988.
13. **Helenius, A. and Simons, K.**, *Biochem. Biophys. Acta*, 415, 29, 1975.
14. **Tanford, C.**, *The Hydrophobic Effect*, J. Wiley & Sons, New York, 1980.
15. **Helenius, A., McCaslin, D. R., Fries, E., and Tanford, C.**, *Methods Enzymol.*, 56, 734, 1979.
16. **Rosen, M. J.**, *Surfactants and Interfacial Phenomena*, John Wiley & Sons, New York, 1978.
17. **Tanford, C.**, *The Hydrophobic Effect*, John Wiley & Sons, New York, 1980.
18. **Wennerstrom, H. and Lindman, B.**, *Phys. Rep.*, 52, 1, 1979.
19. **Zulauf, M.**, *Physics of Amphiphiles: Micelles, Vesicles and Microemulsions*, Degiorgio, V. and Corti, M., Eds., Elsevier, Amsterdam, 1984.
20. **Weckstrom, K.**, *FEBS Lett.*, 192, 220, 1985.
21. **Zulauf, M. and Rosenbuch, J. P.**, *J. Phys. Chem.*, 87, 856, 1983.
22. **McPherson, A.**, *Preparation and Analysis of Protein Crystals*, John Wiley & Sons, New York, 1982.
23. **Mitchell, D. J., Tiddy, G. J. T., Waring, L., Bostock, T., and McDonald, M. P. J.**, *Chem. Soc. Faraday Trans.*, 79, 975, 1983.
24. **Evans, D. F. and Ninham, B. W.**, *J. Phys. Chem.*, 90, 226, 1986.
25. **Tiddy, G. J. T.**, *Modern Trends in Colloidal Science*, Eicke, H. F., Ed., Birkhauser Verlag, Basel, 1985, 148.
26. **Benjamin, L.**, *J. Colloid Interface Sci.*, 22, 386, 1966.
27. **Hermann, K. W. and Benjamin, L.**, *J. Colloid Interface Sci.*, 43, 485, 1973.
28. **Becher, P. and Triffilletti, S. E.**, *J. Colloid Interface Sci.*, 23, 478, 1966.
29. **Le Maire, M., Kwee, S., Andersen, J. P., and Moller, J. V.**, *Eur. J. Biochem.*, 129, 525, 1983.
30. **Foster, D. L. and Fillingame, R. H.**, *J. Biol. Chem.*, 254, 8230, 1979.
31. **Dean, W. L. and Tanford, C.**, *Biochemistry*, 17, 1683, 1978.
32. **Bordier, C.**, *J. Biol. Chem.*, 256, 1604, 1981.
33. **Giege, R., Dock, A. C., Kern, D., Lorber, B., Thierry, J. C., and Moras, D.**, *J. Crystal Growth*, 76, 554, 1986.
34. **Markovic-Housley, Z. and Garavito, R. M.**, *Biochem. Biophys. Acta*, 869, 158, 1986.
35. **Lugtenberg, B. and Van Alphen, L.**, *Biochem. Biophys. Acta*, 737, 51, 1983.
36. **Michel, H.**, *EMBO J.*, 1, 1267, 1982.
37. **Rosevear, P. B., VanAken, T., Baxter, J., and Ferguson-Miller, S.**, *Biochemistry*, 19, 4108, 1980.
38. **Pace-Asiak, C. R. and Smith, W. C.**, *Enzyme.*, 16, 543, 1983.
39. **Van Der Ouderaa, F. J. G. and Buytenhek, M.**, *Methods Enzymol.*, 86, 60, 1982.
40. **Roth, G. J., Siok, C. J., and Ozols, J.**, *J. Biol. Chem.*, 255, 1301, 1980.
41. **Merlie, J. P., Fagan, D., Mudd, J., and Needleman, P.**, *J. Biol. Chem.*, 263, 3550, 1988.
42. **Malkin, R.**, *Ann. Rev. Plant Physiol.*, 33, 455, 1982.
43. **De Grip, W. J.**, *Methods Enzymol.*, 81, 256, 1982.
44. **Picot, D. and Garavito, R. M.**, *Cytochrome P-450: Biochemistry and Biophysics*, Schuster, I., Ed., Taylor and Francis, Philadelphia, 1989, 29.

Chapter 5

PROTEIN-DETERGENT MICELLAR SOLUTIONS FOR THE CRYSTALLIZATION OF MEMBRANE PROTEINS: SOME GENERAL APPROACHES AND EXPERIENCES WITH THE CRYSTALLIZATION OF PIGMENT-PROTEIN COMPLEXES FROM PURPLE BACTERIA

Wolfram Welte and Thomas Wacker

TABLE OF CONTENTS

I. INTRODUCTION

Usually, a protein crystallizes only if a precise structure is common to all copies of the protein present in the mother solution.[1] Thus, the following sources of inhomogeneities must be avoided:

1. Contamination by other proteins, therefore high purity is required
2. Inhomogeneous oligomerization of the protein, therefore a single type of oligomer must be present
3. Denaturation, therefore parameters like pH, ionic strength, the type of detergent used for solubilization, composition, etc., have to be selected so as to stabilize the structure of the protein as much as possible

Solubilization of the protein with a detergent is necessary in order to prepare an aqueous solution of a single type of oligomer, as membrane proteins in the absence of detergents would form aggregates. However, the environment of a micelle differs from a lipid bilayer. Some membrane proteins can be solubilized with full activity in almost every nonionic detergent. Others are active after solubilization only in specific detergents. The substitution of the lipid bilayer by a micelle thus can cause partial or complete loss of the native structure of the protein and concomitant inactivation. In the course of our work we were not always able to find a suitable detergent for a given protein. We therefore devoted part of our work to the search for new detergents capable of conserving and stabilizing the structure of labile membrane protein complexes. When we started to try to solubilize proteins with amphiphiles, we first had to learn which properties of an amphiphile are essential for binding to occur to a membrane protein. Section II of this review will deal with this aspect. In Section III we will report our observations with the stability of two labile membrane proteins (the Ca^{++}-ATPase of sarcoplasmatic reticulum and rhodopsin of bovine rod outer segments) after solubilization with different detergents. Section IV will summarize experiences with different purification methods and Section V will deal with crystallization techniques.

II. SEARCHING FOR NEW DETERGENTS

We tested several amphiphiles for their ability to solubilize the B800-850 light-harvesting complex from the purple bacterium *Rhodopseudomonas (Rp.) palustris*. The test makes use of the extraordinary stability of this complex, which can be dialyzed or adsorbed to a DEAE-column and washed with buffer without a detergent.

The amphiphile to be tested for its ability of solubilizing the protein was then added to a concentration of about three times its critical micellar concentration (CMC) to the buffer used. If the protein was eluted, it was concluded that the substance forms mixed protein-detergent micelles and is thus useful as a detergent. Our results are summarized in the remainder of this section.

A. ALKYLGLUCONAMIDES

Alkylgluconamides[2] were insoluble at room temperature. Upon heating, the crystals dissolved at 80°C. After cooling to below 80°C, the solution became turbid and a gel-like solid formed from which crystals grew within 1 to 3 d of incubation at room temperature.[3] Crystals of N-(n-octyl)-D-gluconamide (CH_3-$(CH_2)_7$-NH-CO-$(CHOH)_4$-CH_2OH) were analyzed by X-ray structure analysis.[4] The structure shows a system of hydrogen bonds between hydroxyl groups, carboxyl groups, and amide groups of neighboring gluconamide moieties, as shown in Figure 1. Evidently, below 80°C the interaction of the amphiphiles is dominated by intermolecular hydrogen bonds, which cannot be broken by a purely hydrophobically interacting protein.

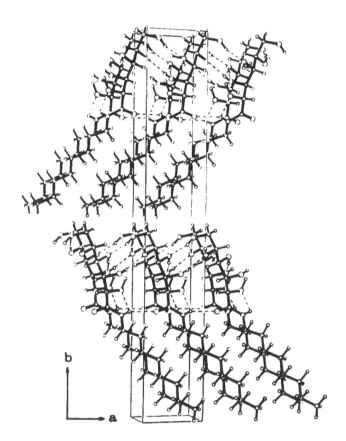

FIGURE 1. Packing of *N*-(*n*-octyl)-D-gluconamide in the unit cell. Hydrogen bonding interactions are drawn as dashed lines, atoms H, C, N, O are indicated with increasing radii. (Adapted from Zabel, V., Müller-Fahrnow, A., Hilgenfeld, R., et al., *Chem. Phys. Lip.*, 39, 313, 1986.)

On the other hand, octanoyl-*N*-methyl-glucamide, which has a closely related structure, $(CH_3-(CH_2)_7-CO-N(CH_3)-CH_2-(CHOH)_4-CH_2-OH)$ is soluble at room temperature and solubilizes proteins.[5] The methyl group, attached to the amide-nitrogen, apparently prevents the formation of a hydrogen bonding system. The presence of the methyl group could either prevent the participation of the amide bond in a hydrogen bonding system, or it could sterically prevent close approach of neighboring hydrocarbon chains necessary for intermolecular hydrogen bonding to occur.

We suppose, that a similar situation exists in the pair of α- and β-ω-D-octylglucopyranoside. The high melting temperature T_c of the α-, as compared to the β-stereoisomer,[6] indicates that the α-isomer has a much more favorable molecular geometry for the formation of an intermolecular hydrogen bonding system, as compared to the β isomer.

Hydrogen bonding between adjacent carbohydrate moieties has indeed been found in crystals of heptylmannopyranoside and xylopyranoside, octyl- and decyl-α-glucopyranoside,[7] and also in decyl-*N*-methylglucamide.[8]

B. ETHER- AND ESTERBIPHENYLOCTAOXYETHYLENE
All ΦΦEO$_9$:

$$(CH_2=CH-CH_2-O-\bigcirc-\bigcirc-O-(CH_2-CH_2-O)_9-CH_3)$$

$$(CH_3=CH-CH_2-O-\bigcirc-\bigcirc-O-CH_2-\underset{\underset{O}{\|}}{C}-O-(CH_2-CH_2-O)_8-CH_3)$$

were a kind gift from Dr. B. Lühmann and synthesized according to Schleicher et al.[9,10] Both were water-soluble at room temperature with a CMC of 260 and 190 μM, respectively.[9] However, only the latter substance is capable of eluting the B800-850 complex from the DEAE column. We presume that in All$\Phi\Phi$EO$_9$ the biphenyl groups of neighboring amphiphiles stick together due to a dispersion interaction caused by the anisotropy of the polarizability of the biphenyl group.[11] We presume further that this interaction is prevented in All$\Phi\Phi$AcEO$_8$ by the hydrated ester-carboxyl group.

C. CONCLUSIONS

When searching for a new detergent, it has to be taken into consideration that detergents usually bind to proteins by hydrophobic interaction, as the half-saturation concentration of detergent-to-protein binding is generally close to the CMC of the detergent.[6] Thus hydrophobic binding to a protein is possible only if there are no additional intermolecular interactions of the amphiphiles. We interpret the three previously mentioned pairs of amphiphiles as each consisting of a nonbinding form, in which additional intermolecular interactions of the amphiphiles occur, and a binding form, in which such additional interaction are suppressed either by a bulky side group or by an altered stereoconfiguration.

III. STABILITY OF MEMBRANE PROTEINS IN MICELLAR SOLUTION

A. THE CA^{++}-ATPASE OF SARCOPLASMIC RETICULUM

Ca^{++}-ATPases pump Ca^{++} against a concentration gradient by use of the free energy of hydrolysis of ATP. The ATPase cycles between two conformational states. In state E, a high affinity binding site for ATP and two high affinity binding sites for Ca^{++} exist, all three located on the cytoplasmic domain of the protein. In state E, the protein possesses two low affinity binding sites for Ca^{++} on the luminal side.[12] The protein consists of a single polypeptide chain of M$_R$ = 110,000; the primary structure is known and has been used for a structural prediction.[13] The hydrophilic domain on the cytoplasmic side contains half of the molecular weight of the protein; the remainder lies within the membrane. The ATPase occurs mainly in muscle cells, but it is also found in the endoplasmic reticulum of liver cells[14] and in the plasma membrane of prokaryotes.[15] The Ca^{++}-ATPase is belonging to the P-type ATPases.[16] Two dimensional crystals have been prepared and analyzed by image filtering.[17,18] We will present a brief summary of our experiments; a more detailed report has been given.[54]

A vesicular preparation of the Ca^{++}-ATPase was supplied by Dr. Colin Restall, Royal Free Hospital, London, U.K. The vesicles were prepared from rabbit hind leg muscle following the method by Nakamura et al.[19]

The ATPase activity was used as an assay for the structural integrity of the protein, and was determined by monitoring the oxidation of NADH in a coupled enzyme assay according to Madden et al.,[20] A detergent was added to a desired concentration to the test medium in order to investigate its interaction with the protein. Octyl-, decyl-, dodecyl-, and myristoyl-derivatives of 1-alkyl, 3-phosphorylcholine propanediol (2-deoxylysolecithin, or ES-n-H with n = 8, 10, and 12) were a kind gift of Dr. H. U. Weltzien, Max-Planck Inst. für Immunbiologie Freiburg and were synthesized as previously described.[21]

Figure 2 shows the typical result obtained with the following detergents: β-ω-D-octyl-glucopyranoside, octyltetraoxyethylene, LDAO, and octanoyl-*N*-methyl glucamide. The pro-

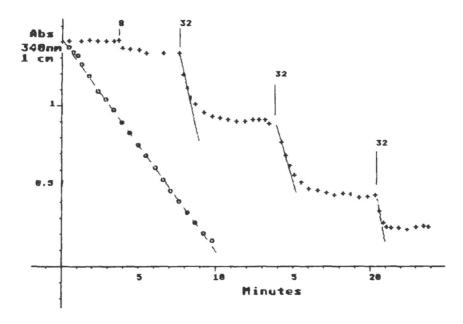

FIGURE 2.   ATPase activity with (o) 0.45% w/v and (+) 0.6% w/v of octylglucoside (CMC = 0.6%) in the test medium. At the times indicated by a vertical bar, 8, 32, 32, and 32 μg of the ATPase were added to the test medium containing 0.6% of the detergent. It will be noted that the ATPase was quickly inactivated at this detergent concentraiton. At 0.45% of octylglucoside (below the CMC), and with 8 μg ATPase, no inactivation occurred within the 10 min observation.

tein was irreversibly inactivated in the presence of these detergents at concentrations corresponding to three times the CMC.

For the ES-n-H series, the activity was determined as a function of detergent concentration. The results are shown in Figure 3. The octyl- and the decyl derivatives failed to preserve activity at concentrations above the CMC, as was the case with the group of detergents of Figure 2. With longer chain derivatives, however, the ability of the deoxylysolecithins to preserve the activity of the ATPase varied strongly with the alkyl chain length. The lauryl derivative was the shortest analogue capable of solubilizing the ATPase in active form. However, the presence of a hydrocarbon chain length of 12 C-atoms alone is not sufficient for a detergent to solubilize the ATPase in active form, as can be seen from the experiment with LDAO which inactivated the ATPase.

Other authors use dodecyloctaoxyethylene ($C_{12}E_8$) for solubilizing the active enzyme,[22-24] They find no residual lipid in the mixed protein-$C_{12}E_8$ micelles. In our experiments, C_8E_4 failed to preserve the ATPase activity.

These findings indicate that both a hydrophobic tail of at least 12 C-atoms and a suitable polar head group (e.g., a phosphorylcholine or a polyoxyethylene group) are prerequisites for a detergent to preserve the activity of the ATPase.

B. RHODOPSIN FROM BOVINE ROD OUTER SEGMENTS

Rhodopsin is the light receptor protein involved in the primary photochemical reactions of vision, containing 11-*cis* retinal as chromophore. It is the major integral protein of the disc membrane stacks in the outer segment of rod cells. Its molecular weight is 39,000, and secondary structure predictions have been made on the basis of the sequence.[25] The polypeptide chain is predicted to form seven transmembrane helices, connected to each other by hydrophilic loops.

After excitation, the protein relaxes through a series of spectrally detectable intermediates into a temperature- and pH-dependent equilibrium of metarhodopsin I (MI) and metarho-

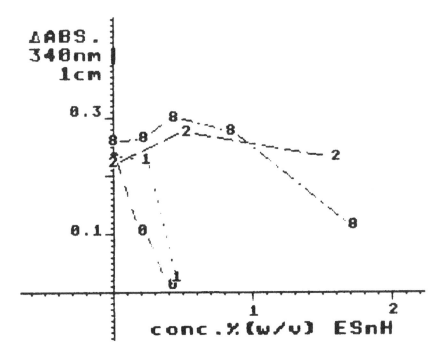

FIGURE 3.   ATPase activity as a function of concentration of the octyl-(8), decyl-(0), undecyl-(1) and dodecyl-(2) derivative of desoxylysolecithin. The CMC are 1.8, 0.17, 0.02, and 0.016% (w/v), respectively.

dopsin II(MII).[26] The two spectral states MI and MII are thought to correspond to two conformational states. The MI ↔ MII equilibrium can be shifted to MI by pressure[27] or by a liquid-to-gel state transition of the lipid environment.[28] The latter authors have proposed a model explaining the influence of the lipid hydrocarbon environment through the exertion of surface tension on the membrane embedded surface of rhodopsin, resulting in a radial pressure on the protein. According to their model, a high degree of fluidity in the protein environment favors the solvation of the protein surface, thereby reducing the interfacial tension at the protein-lipid interface. A low degree of fluidity, as provided by a gel-state hydrocarbon phase, thus could impose sufficient pressure to shift the equilibrium towards MI.

Previous studies with solubilized rhodopsin used detergents with their hydrophobic function provided by alkyl chains (e.g., LDAO, octylglucoside, dodecylmaltoside), or a rigid steroid ring system (e.g., digitonin). While with all the protein-detergent micelles of the first group the equilibrium was shifted to MII, the rhodopsin-digitonin micelle showed an almost normal MI ↔ MII equilibrium. We wished to determine whether the hypothesis of Baldwin and Hubbell, deduced from lipid bilayer data, could explain these observations as well. A brief summary of our results follows: for details see Schleicher et al.[9] The methods for synthesis of AllΦΦEO$_8$ and for preparation of washed disc membranes and rhodopsin have been described.[9] MII formation was measured by comparing the light-induced changes in the difference between absorption at 380 and 417 nm. The fraction (f) of photoexcited rhodopsin which assumes the MII conformation can be calculated by use of the extinction coefficient for MII formation.[29]

Table 1 summarizes the results. AllΦΦAcEO$_8$-rhodopsin micelles showed a MI↔MII equilibrium similar to that of rhodopsin in washed membranes. In contrast, ES-12-H-rhodopsin mixed micelles showed the shift to the MII conformation that other authors had observed previously with all detergents having an alkyl chain as hydrophobic part. This

TABLE 1
Fraction of Photoexcited Rhodopsin Which Assumes the MII Conformation in Washed Disc-Membrane Preparations, in ES12H, and in AllΦΦAcEO$_8$

	Washed membranes	ES12H	AllΦΦEO$_8$
pH 6, 21°C	0.95	1.0	0.62
pH 7.5, 8°C	0.31	1.0	0.14

Note: The fraction in the latter was similar to that in washed membranes for two standard conditions of pH and temperature. In ES12H, the equilibrium was shifted to the MII conformation.

seems to be in contradiction to the work of Baldwin and Hubbel, who found inhibition of MII when rhodopsin was reconstituted in bilayers of saturated lipids with alkyl chain lengths shorter than 14 C-atoms. This apparent conflict can be resolved by considering that the interfacial tension is not only determined by molecular structure but also by packing constraints. Conceivably, low curvature of the bilayer could favor the alignment of extended hydrocarbon chains and result in poor solvation and high surface tension. On the other hand, high curvature of micelles would impose packing constraints favoring a disordered fluid phase which would solubilize the protein surface well, resulting in low surface tension.

Rhodopsin in digitonin and in AllΦΦAcEO$_8$ shows a similar MI $\leftrightarrow$ MII equilibrium to that in the native bilayer environment. We thus believe that a sterol ring system and the biphenyl group impose rigidity to the hydrophobic core of the micelle, thus providing the interfacial tension which is necessary for the native free energy gap between MI and MII. The results with the biphenyl detergent demonstrates that a complete sterol ring system is not necessary for this equilibrium.

The interfacial tension between a belt of biphenyl detergents and a hydrophobic protein surface could also be due to the aforementioned interactions of the biphenyl group (see Section II). A protein has to compete with these interactions in order to be solubilized. For a detergent molecule, the interaction with the hydrophobic surface of a protein will represent an "unfavorite interaction" as compared to an interaction with a shell of neighboring biphenyl detergent molecules.

It is possible that many labile membrane proteins consist of poorly interacting transmembrane α-helices which retain their native tertiary structure only when put under a certain interfacial pressure. The development of new detergents with a rigid but small biphenyl group in their hydrophobic domain could be of importance for the purification and crystallization of labile membrane protein complexes.

IV. CHROMATOGRAPHIC PURIFICATION

When a detergent which preserves the activity of the protein has been found chromatographic procedures have to be used to prepare a homogeneous solution as described in Section I.

We tend to use chromatofocusing as final purification step. As the pH of elution depends on the distribution of amino acids on the hydrophilic surface of the protein-detergent micelle, we consider this method to have the capacity of separating partially denatured material of the desired protein as well as impurities. Chromatofocusing separates proteins according to their isoelectric points, as does isoelectric focusing which has been shown to be a beneficial final purification step before crystallization[30]; but chromatofocusing has the further advantage that it does not require the application of an electric field which may cause damage to the protein.

A. THE B800-850 COMPLEX FROM *RHODOBACTER CAPSULATUS*

The B800-850 complex of *Rhodobacter (Rb.) capsulatus* was purified as described by Welte et al.[31] Crystallization yielded triclinic and orthorhombic crystals. The same polypeptide pattern was found for both types of crystals in SDS-Polyacrylamide gel electrophoresis (SDS-PAGE). We subjected the preparation to a further chromatographic purification by chromatofocusing in the presence of 0.04% LDAO. The crystals grown from the eluate were exclusively orthorhombic. We infer that chromatofocusing removes residual lipid from the protein, as we could not detect any difference in the protein pattern of the final as compared to the initial material.

B. THE REACTION CENTER(RC)-B875 COMPLEX FROM *RHODOBACTER PALUSTRIS*

The complex was purified as described by Wacker et al.[32]. In chromatofocusing, three bands eluted at pH 7.2 (peak 1), 6.9 (peak 2), and 6.4 (peak 3). When the protein was washed with 1 column volume of the starting buffer plus 0.7% LDAO after loading the column, peak 1 and in some cases peaks 1 and 2, were missing. The spectra of peaks 1 and 2 were similar. Typical spectra of peaks 2 and 3 are shown in Figure 4A. The difference between the two spectra (Figure 4B) is characteristic of the B800-850 complex of *Rhodobacter(Rp.) palustris*. Pure B800-850 complex elutes at pH 7.0. We conclude that peak 1 and peak 2 consist of a RC-B875-B800-850 complex possible held together by residual lipid. Washing with LDAO presumably extracts the lipid and desorbs the B800-850 complex from the column. The best yield of crystals was obtained from peak 3 material.

C. MISCELLANEOUS OBSERVATIONS

We use gel filtration on Superose 6B to test our final preparations for homogeneity of size (oligomerization). With more than 50 μM Ca^{++} in the elution buffer, the Ca^{++}-ATPase tends to elute as a mixture of monomers and dimers, while in the absence of Ca^{++} it is monomeric.

Addition of glycerol to the elution buffer helped, in many cases, to preserve the activity of the protein during chromatographic runs. We routinely add 15% glycerol in the final chromatographic run for purifying the B800-850, the RC-B875 complex and the Ca^{++}-ATPase. It is known that substances like sugars and glycerol stabilize proteins by an unfavorable energy of interaction.[33,34] The unfavorable energy of interaction leads to an exclusion of glycerol from the protein surface[35] which results in an interfacial pressure.[33] This mechanism of stabilization may be similar as the one mentioned in Section III.B.

When glycerol was omitted, the yield of crystals of the B800-850 complex of *Rb. capsulatus* was smaller and the color of the crystals frequently turned from red to green. This color change is indicative of a loss of carotenoid. It is known that B800-850 complexes are stable at alkaline pH (8.0 to 8.5), but labile at pH 7.0, their pH of elution in chromatofocusing. In addition, loss of carotenoids is known to destabilize the B800-850 complex. We take it that upon chromatofocusing in the absence of glycerol, the complex either loses part of its carotenoids or possibly some tightly bound lipid which serves to stabilize the complex.

The addition of 2 to 4 mM glutathione was necessary in order to preserve the activity of the Ca^{++}-ATPase during gel filtration as well as during storage at room temperature. We presume that glutathione protects reactive cysteinyl groups on the hydrophilic domain of the Ca^{++}-ATPase.[36]

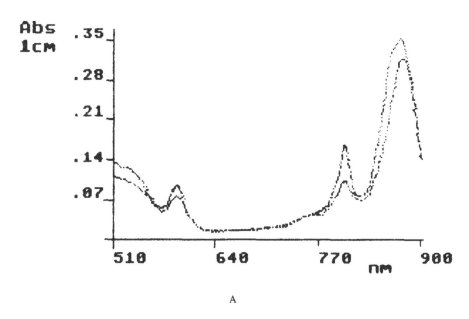

A

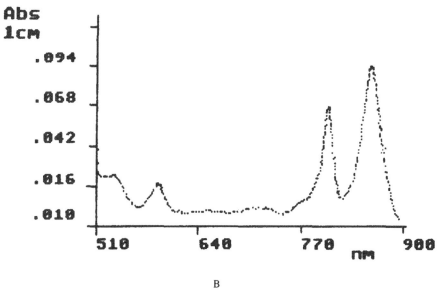

B

FIGURE 4. (A) Spectra of material eluting in peak 1 or 2 (upper spectrum) and of peak 3 (lower spectrum). (B) Difference obtained after subtracting the peak 3 spectrum from the peak 1 or 2 spectrum. The difference is characteristic for the B800-850 complex.

V. CRYSTALLIZATION

Probably, all membrane protein crystals obtained so far represent a lattice formed by mixed protein-detergent micelles.[37] The behavior of micelles in aqueous solutions can be expected to play a role in crystal formation. For its description see Chapter 2 and References 38 to 45.

A. MATERIALS AND METHODS

Prior to crystallization attempts the protein-detergent solution was concentrated to a protein concentration of 20 mg/ml with an ultrafiltration cell (Amicon). We used membranes with a cut-off limit as large as possible (XM50 or XM100, Amicon) in order to prevent, as far as possible, concentration of the pure detergent micelles. The protein then was dialyzed overnight against a buffer with low ionic strength (usually 10 mM and the desired detergent concentration. This was to make sure that the elevated detergent concentration due to the ultrafiltration process, was reduced to the concentration desired for crystallization. We ordinarily use a concentration of detergent equivalent to twice the CMC. After dialysis, the protein concentration was adjusted to 10 mg/ml. For presaturation, the protein solution was mixed with an equal volume of "presaturation solution" containing buffer with "high ionic strength" (100 mM), but no detergent and twice the desired final concentration of precipitant (PEG, ammonium sulphate, or phosphate). This also adjusts the pH of the mixture to the desired value due to the high buffering capacity of the presaturation solution. Small amphiphiles or other additives were usually added to the presaturated protein solutions from concentrated stock solutions. For crystallization two methods were tried.

1. Vapor Phase Equilibration of the Presaturated Protein Solution With a Reservoir Buffer

The latter was prepared by dilution of the presaturation solution with pure buffer to the desired final PEG/ammoniumsulfate/phosphate concentration. Ten microliters of the presaturated protein solution were put in a depression of a microtitre plate (Greiner and Söhne GmbH, FRG, oberflächenbehandelt) and 1 ml of reservoir buffer was divided between the remaining depressions. The microtitre plate was closed with the cover, sealed with silicon grease and parafilm, and incubated at room temperature or 4°C in the dark.

To test the quality of the crystals by X-ray diffraction, usually big crystals are needed. As growing and mounting big crystals can be very tedious, we tried growing them directly in a glass capillary. A capillary held vertically was filled from the bottom to about one third of its length with the presaturated protein solution. No sealing at the top was necessary because of the long, empty capillary, which prevents a quick evaporation of water. After some 5 d, all of the protein had crystallized. The capillary was sealed and centrifuged at low speed, to pellet all crystals. The X-ray photograph shows a typical mosaic pattern. We consider this method to be a useful quick test of the influence of additives on the quality of crystals, when the basic crystallization conditions are known.

2. Microdialysis

In our dialysis experiments we used Spectrapor membranes with a cutoff limit of 3500 mol wt.

The reservoir solution has to contain all the substances which are to be present in the protein solution during crystallization, i.e., detergents and small amphiphiles have to be added. However, as a precipitating agent only polyethyleneglycol is possible as the salts would quickly equilibrate between both compartments. A microdialysis chamber similar to the one proposed by Zeppezauer was used.[46] Later, a new cell was constructed which is shown in Figure 5.

The protein solutions in the microtitre plates were observed under a microscope. Crystals were analyzed by X-ray diffraction with a rotating anode X-ray source (Siemens AG, FRG, operated at 40 kV/250 mA) in a precession camera (Reciprocal lattice explorer, STOE, Darmstadt, FRG) or by spectroscopy with polarized light in a homemade microspectrophotometer.

B. RESULTS AND DISCUSSION
1. Vapor Phase Equilibration Versus Microdialysis

In vapor phase equilibration, equilibration of the protein solution with the reservoir was attained after approximately 2 d. The process could be slowed down by using larger volumes of protein solution. The geometry of the depression, i.e., the ratio of surface-to-volume of

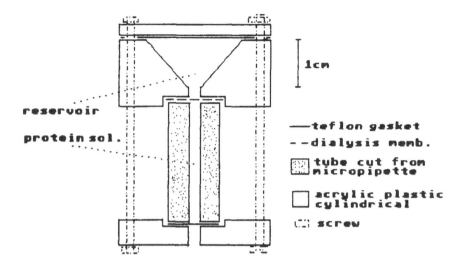

FIGURE 5. Scheme of a microdialysis cell which was designed to prevent quick diffusion of water from the protein solution to the reservoir by keeping the volume of the protein solution constant. The crossection is not in correct scale, the protein volume was approximately 20 μl.

the protein solution, seemed to influence the speed of equilibration. When the vacuum grease was omitted, a slow evaporation of water out of the microtitre plate occurred, which could be used to slowly cross the solubility limit of the protein in order to grow large crystals within approximately 2 weeks.

Microdialysis has the advantage of keeping the concentration of detergent and salt in the protein solution constant when increasing the concentration of the precipitating agent. On the other hand there is a quick flow of water from the presaturated protein solution to the reservoir. Equilibration in the case of 10 to 20 μl volumes of protein solution takes place within less than one night, so that equilibration is faster in the latter method. This quick equilibration was disadvantageous for the proteins with which we work.

We tried to inhibit the water flow by constructing the cell shown in Figure 5. The concept was to keep the volume of the protein solution constant by sealing the bottom of the glass cylinder with a teflon sheet and the top with the dialysis membrane. We expected a slow equilibration process within several days due to the slow passage of PEG across the dialysis membrane. However, after 1 or 2 d air bubbles usually formed within the protein solution, probably because of the osmotically reduced pressure in this compartment.

Another method to slow down the equilibration process was described by Weber and Goodkin.[47] In their device, the protein is separated from the reservoir not only by a dialysis membrane, but also by a capillary filled with the presaturated solution. Due to the long diffusion distance there is slow equilibration lasting for days or weeks; the technique can also be used for ammonium sulfate. We have not yet tried the method.

All of the following crystallization experiments were made with the vapor phase equilibration method.

2. Survey of phenomena observed in crystallization experiments

Phase separations into a detergent poor phase and a detergent enriched phase with and without liquid crystalline order can be expected from the schematic phase diagram in Figure 5, as well as pure crystal growth. Usually, the protein partitions into the detergent enriched phase. The following plates show that all these phenomena were indeed observed in crystallization experiments with B800-850 complexes from *Rp. palustris*. Plate 1A* shows a

* Plate 1 appears after page 120.

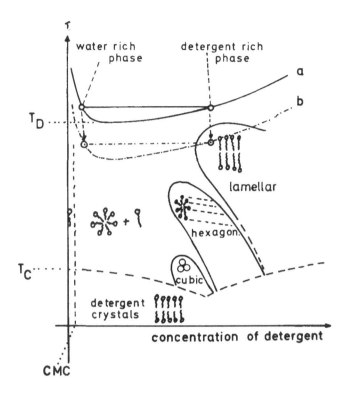

FIGURE 6. Schematic-phase diagram for a nonionic detergent water system. The abscissa and ordinate are without scale. Liquid crystalline phases are typically found at detergent concentrations greater than about 30%. Inserted in the liquid crystalline areas are schemes of the liquid crystalline aggregates. More detailed pictures are given in Reference 37. The phase separation boundary for the system without PEG is labeled "a". The phase separation boundary downshifted by PEG with its detergent-rich phase in the lamellar liquid crystalline area is labeled "b".

two-phase system with an isotropic detergent-rich phase colored by the carotenoid of the B800-850 complex. Plate 1B shows a two-phase system with a hexagonal liquid crystalline texture in the detergent-rich phase. Plate 1C shows a two-phase system with a hypothetically lamellar liquid crystalline detergent-rich phase and crystals growing on the interface. Systems showing crystals alone were also found.

Micelle interaction phenomena thus are found in mixed protein-detergent-PEG solutions. In order to estimate their importance for the crystallization, it is convenient to imagine two limiting forms of protein-detergent micelles. One extreme is a micelle with a major part of the surface formed by the polar head groups of detergent molecules and therefore with poor protein-specific interactions. The protein hydrophilic surface in this case need not necessarily be small. It may also be that detergents bind to the hydrophilic surfaces in the presence of PEG or ammonium sulfate.[43] This assumption is further supported by the observed effect of detergents upon crystallization of water-soluble proteins.[48]

The other extreme would be a protein-detergent micelle where the major part of the surface is formed by hydrophilic protein surfaces. The latter protein should crystallize in a similar manner as a water soluble protein. The hydrophilic protein surfaces by their interaction should form the lattice. The detergent belt should play a subordinate role at the crystallization conditions. Its interaction should neither impede crystal formation, nor sterically impede the formation of a crystal lattice. For crystallization of the former type of membrane protein one has no other choice than to use conditions close to, or even within the phase separation region, where aggregation of micelles is favored. The formation of the crystal would be influenced by the arrangement of the detergent around the protein, its tendency for liquid crystalline ordering, and so forth, all affecting protein interactions. Crystal habits would change upon use of other detergents and temperature.

3. Attempts to Avoid Phase Separations

Crystallization of proteins out of phase separations has several disadvantages. The protein may be denatured by the high detergent concentration. Every droplet may induce crystallization, so that there is a high yield of small crystals as shown in Plate 1C. During mounting of the crystal in a capillary, the contact with the detergent-rich phase has to be avoided, as this may cause irreversible damage to the crystal order.[49] We therefore attempt to avoid phase separations and we believe that this strategy is equivalent to strengthening the interactions of the hydrophilic protein surfaces. We tested different detergents and additives as well as ionic strength and temperature.

Phase separations may be driven by hydrophobic aggregation of detergent micelles, by unfavorable energies of interaction with PEG, but also by unfavorable interactions of PEG with protein (salting out effects).[44,50] In order to see if phase separation is exclusivey driven by hydrophoobic and by PEG-induced detergent micelle aggregation, we prepared aqueous buffer solutions, containing all substances except for the protein at concentrations typical for the crystallization of the B800-850 complex from *Rb. capsulatus*. Figure 7 shows that at a concentration of 21 to 24% (w/w) PEG the temperature at which phase separation was observed was well above room temperature. If protein was present at 5 mg/ml but at otherwise identical conditions, red phase separation droplets were observed, showing that the protein had segregated into a separate phase. Thus, the energy of interaction of PEG with mixed protein-LDAO micelles is more unfavorable than with pure LDAO micelles, so that the presence of the protein induces phase separation and liquid crystalline ordering. The hypothesis is supported also by the facts that crystallization and phase separation occur at the same PEG concentration (see Figure 8) and that the same phenomena are observed at 4°C.

These observations show, that phase separation is not driven exclusively by a temperature dependent hydrophobic effect or by an unfavorable interaction of PEG with pure detergent micelles. An additional contribution is an unfavorable interaction of the hydrophilic protein surfaces with PEG. PEG in general has an unfavorable interaction with proteins[50] which seems to be related to its salting out effect.[51]

Phase separations in aqueous solutions of dextrans and polyethylenglycols have been investigated by Albertsson.[52] A protein-detergent micelle might be considered in a rough approximation as a polymer molecule. Indeed, phase separation and crystallization of membrane proteins in PEG solutions has similarities with the dextran-PEG system. Upon increase of either the molecular weight of PEG or of the protein-detergent micelle, the system approaches towards a phase boundary or crystallization. This can be rationalized by an unfavorable interaction energy and the decrease of the entropy of mixing with increasing molecular weight.[50] If phase separations are observed, we therefore try to avoid them by using PEGs of lower molecular weight.

Another method to avoid phase separations is to add short chain dimethylamineoxides as small amphiphiles. In this manner, the phase separation could be shifted to higher PEG concentrations and even made to disappear by using heptylDAO or heptanetriol as additives (see Figure 8).

At higher temperatures, the crystals became more and more unstable and disappeared at about 28°C by forming droplets. The same was seen occasionally after raising the concentration of PEG. We interpret this melting of the crystals as signifying that hydrophobic micelle interactions become predominant over protein interactions.

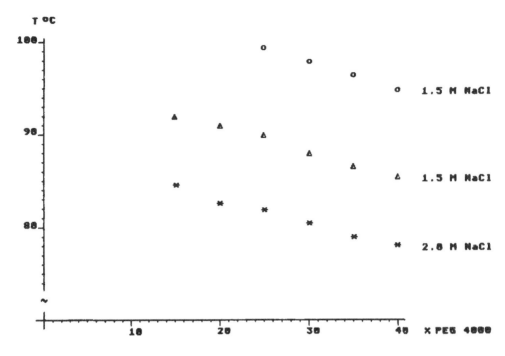

FIGURE 7.   Temperature of phase separation in 25 m*M* Tris buffer, pH 8 containing 0.8% w/v LDAO and concentrations of PEG 4000 and sodium chloride as indicated in the figure.

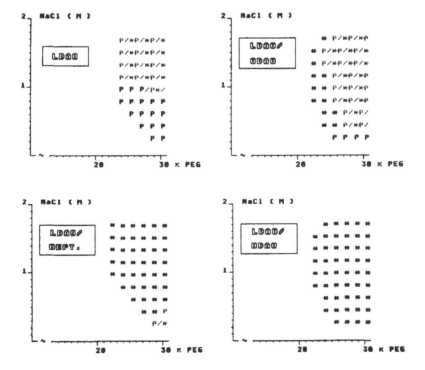

FIGURE 8.   Phase diagrams for crystallization of the B800-850 complex of *Rb. capsulatus* in the presence of different small amphiphiles. The buffer contained 25 m*M* Tris-Cl pH 8 and 0.08% w/v LDAO. Small amphiphiles were octyldimethylamineoxide (ODAO) at 3% w/v, heptyldimethylamineoxide (HDAO) at 5% w/v and heptane-triol (HEPT) at 3% w/v. Symbols: *, B800-850 crystals; P, phase separation; and P/*, phase separation and badly formed crystals.

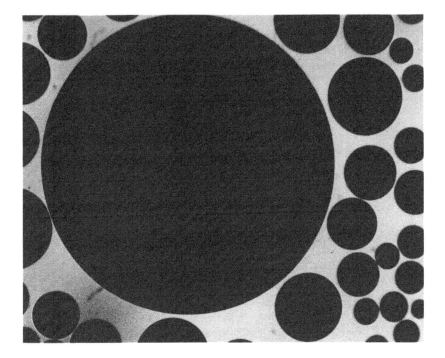

A

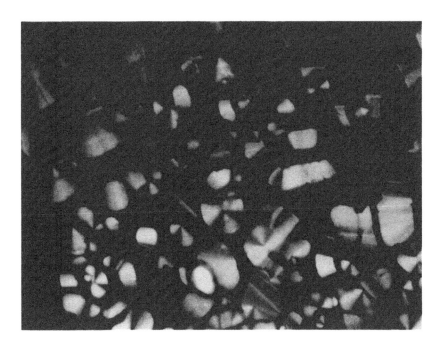

B

Plate 1. Phase separations photographed between crossed polarizers observed in aqueous solutions containing PEG, detergent and the B800-850 complex. The latter is colored by its pigments. (A) An isotropic detergent-rich phase. (B) A detergent-rich phase showing a texture typical for hexagonal liquid crystalline ordering. (C) A detergent-rich phase with a texture tentatively assigned to a lamellar liquid crystalline ordering. Crystals are seen near the interface of the two phases.

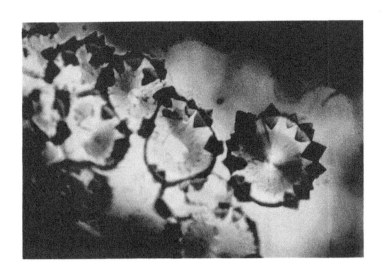

Plate 1C.

It follows that precautions must be taken when mounting crystals in capillaries. The reservoir in the capillary should be identical to the mother solution and the temperature should be low (e.g., 10°C) during exposure to the X-ray beam. The addition of sodium chloride was necessary for obtaining well-shaped crystal. Similar phase diagrams were found with $MgCl_2$ at lower concentrations. We think that the presence of salt favors the interaction of the proteins.

The crystals of the reaction center-B875 complex were more labile. They melted to form droplets upon mounting into capillaries. When the crystallization was done in the capillary, as described above, the powder diagram showed a few reflections to a resolution of approximately 30 Å. The biggest crystals of this protein (approximately 100 μ × 100 μ × 100 μ) were obtained with 1-dodecyl propanediol-3-phosphorylcholine as detergent[21] and heptanetriol as a small amphiphile, in 25 mM Tris without the addition of salt.

The mode of action of the "small amphiphiles" could be to reduce the radius of the micelle "belt" adhering to the protein, thereby removing sterical hindrances for the formation of the crystal lattice. An alternative explanation is that these substances exchange for some LDAO adsorbed to hydrophobic patches within the hydrophilic protein surface or to the boundary region of hydrophilic and hydrophobic surface areas. This would result in a bigger area of the protein surface accessible to interactions. The latter hypothesis is supported by our observations. As mentioned above, the B800-850 complex from *Rp. palustris* has an isoelectric point near 6.8 and is crystallized near pH 8, where it is negatively charged. We used 1,6-hexanediamine as an additive for crystallization; these will be positively charged at pH 8. X-ray diffraction patterns of the resulting crystals showed a resolution of about 7Å in contrast to about 10 Å for crystals without these substances. We consider that the alkyl chain and NH_2 group of the diamines interact with the hydrophilic protein surface and act as a linker between two protein micelles.

The space group of the B800-850 crystals was orthorhombic C222₁.[31,53] At high concentrations of ammonium sulfate, the B800-850 complex from *Rp. palustris* was unstable. However, another complex from *Rhodospirillum salexigens* could be grown in PEG as well as in ammonium sulfate in a tetragonal space group.

ACKNOWLEDGMENTS

We are grateful to Dr. David Moss, Dr. A. Schleicher and Dr. K. P. Hofmann for critical reading of the manuscript. This work was supported by the Deutsche Forschungsgemeinschaft (SFB60).

REFERENCES

1. **McPherson, A.**, *Preparation and Analysis of Protein Crystals*, John Wiley & Sons, New York, 1982, Chap. 3.
2. **Emmerling, W. N. and Pfannemüller, B.**, Membrantrennsschichten aus monomolekularen Filmen von synthetischen Glycolipiden, *Colloid Polym. Sci.*, 261, 677, 1983.
3. **Pfannemüller, B. and Welte, W.**, Amphiphilic properties of synthetic glycolipids based on amide linkages. I. Electronmicroscopic studies on aqueous gels, *Chem. Phys. Lip.*, 37, 227, 1985.
4. **Zabel, V., Müller-Fahrnow, A., Hilgenfeld, R., Saenger, W., Pfannemüller, B., Enkelmann, V., and Welte, W.**, Amphiphilic properties of synthetic glycolipids based on amide linkages. II. Crystal and molecular structure of *N*-(n-octyl-β-ᴅ-gluconamide), an amphiphilic molecule in head-to-tail packing mode, *Chem. Phys. Lip.*, 39, 313, 1986.

5. **Hildreth, J. E. K.**, N-D-gluco-N-methylalkanamide compounds, a new class of nonionic detergents for membrane biochemistry, *Biochem. J.*, 207, 363, 1982.
6. **Grado, M.**, Solubilization and reconstitution of membrane proteins, Ph.D. thesis, University of Basel, Switzerland, 1982.
7. **Jeffrey, G. A.**, Carbohydrate liquid crystals, *Acc. Chem. Res.*, 19, 168, 1986.
8. **Müller-Fahrnow, A., Zabel, V., Steifa, M., and Hilgenfeld, R.**, Molecular and crystal structure of a nonionic detergent: Nonanoyl-N-methylglucamide, *J. Chem. Soc. Chem. Commun.*, p. 1573, 1986.
9. **Schleicher, A., Franke, R., Hofmann, K. P., Finkelmann, H., and Welte, W.**, Desoxylysolecithin and a new biphenyl detergent as solubilizing agents for bovine rhodopsin, *Biochemistry*, 26, 5908, 1987.
10. **Lühmann, B.**, Über den Einfluss einer Polymerfixierung auf das lyotrope Mesophasenverhalten nichtionischer Amphiphile, Ph.D. thesis, TU Clausthal, FRG, 1985.
11. **Maier, W. and Saupe, A.**, Klärpunkt und Anisotropie der molekularen Polarisierbarkeit kristallin-flüssiger Substanzen, *Z. Naturforschg.*, 12a, 668, 1957.
12. **De Meis, L. and Vianna. A. L.**, Energy interconversion by the Ca^{++}-dependent ATPase of the sarcoplasmic reticulum in *Ann. Rev. Biochem.*, 48, 275, 1979.
13. **MacLennan, D. H., Brandl, C. J., Korczak, B., and Green, N. M.**, Amino-acid sequence of a Ca^{++} + Mg^{++}-dependent ATPase from rabbit muscle sarcoplasmic reticulum, deduced from its complementing DNA sequence, *Nature*, 316, 696, 1985.
14. **Spamer, C., Heilmann, C., and Gerok, W.**, Identification of a Ca^{++}-transport ATPase in human liver endoplasmic reticulum. Abstract TU387, *13th Int. Congress Biochemistry*, Amsterdam, The Netherlands, 1985.
15. **Lockau, W. and Pfeffer, S.**, ATP-dependent calcium transport in membrane vesicles of the cyanobacterium *Anabaena variabilis*, *Biochim. Biophys. Acta*, 733, 124, 1983.
16. **Pedersen, P. L. and Carafoli, E.**, Ion motive ATPases. I. Ubiquity, properties, and significance to cell function, *Trends Biochem. Sci.*, 12, 146, 1987.
17. **Dux, L. and Martonosi, A.**, Two-dimensional arrays of proteins in sarcoplasmic reticulum and purified Ca^{++}-ATPase vesicles treated with vanadate, *J. Biol. Chem.*, 258, 2599, 1983.
18. **Dux. L., Taylor, K. A., Ting-Beall, H. P., and Martonosi, A.**, Crystallization of the Ca^{++}-ATPase of sarcoplasmic reticulum by Ca^{++} and lanthanide ions, *J. Biol. Chem.*, 260, 11730, 1985.
19. **Nakamura, H., Jilka, R. L., Boland, R., and Martonosi, A. N.**, Mechanism of ATP-hydrolysis by sarcoplasmic reticulum and the role of phospholipids, *J. Biol. Chem.*, 251, 5414, 1976.
20. **Madden, T. D., Chapman, D., and Quinn, P. J.**, Cholesterol modulated activity of calcium-dependent ATPase of the sarcoplasmic reticulum, *Nature*, 279, 538, 1979.
21. **Weltzien, H. U., Richter, G., and Ferber, E.**, Detergent properties of water soluble choline phosphatides, *J. Biol. Chem.*, 254, 3652, 1979.
22. **Dean, W. L. and Tanford, C.**, Properties of detergent-activated Ca^{++}-ATPase, *Biochemistry*, 17, 1683, 1978.
23. **Andersen, J. P., Jorgensen, P. L., and Møller, J. V.**, Direct demonstration of structural changes in soluble, monomeric Ca^{++}-ATPase associated with Ca^{++} release during the transport cycle, *Proc. Natl. Acad. Sci. USA*, 82, 4573, 1985.
24. **Vilsen, B. and Andersen, J. P.**, Occlusion of Ca^{++} in soluble, monomeric sarcoplasmic reticulum Ca^{++}-ATPase. *Biochim. Biophys. Acta*, 855, 429, 1986.
25. **Ovchinnikov, Y. A., Abdulaev, N. G., Feigina, M. Y., Artamonov, I. D., Zolotarev, A. S., Kostina, M. B., Bogachuk, A. S., Miroshnikov, A. I., Martinov, V. I, and Kudelin, A. B.**, The complete amino acid sequence of visual rhodopsin, *Bioorg. Khim.*, 8, 1011, 1982.
26. **Hoffman, K. P.**, Photoproducts of rhodopsin in the disc membrane, *Photobiochem. Photobiophys.*, 13, 309, 1986.
27. **Lamola, A. A., Yamane, T., and Zipp, A.**, Effects of detergents and high pressures upon the meta-rhodopsin I - metarhodopsin II equilibrium, *Biochemistry*, 13, 738, 1974.
28. **Baldwin, P. A. and Hubbell, W. L.**, Effects of lipid environment of the light-induced conformational changes of rhodopsin, *Biochemistry*, 24, 2624, 1984.
29. **Parkes, J. H. and Liebman, P. A.**, Temperature and pH dependence of the metarhodopsin I-metarhodopsin II kinetics and equilibria in bovine rod disk membrane suspensions, *Biochemistry*, 23, 5054, 1984.
30. **Bott, R. R., Navia, M. A., and Smith, J. L.**, Improving the quality of protein crystals through purification by isoelectric focusing, *J. Biol. Chem.*, 257, 9883, 1982.
31. **Welte, W., Wacker, T., Leis, M., Kreutz, W., Shiozawa, J., Gadón, N., and Drews, G.**, Crystallization of the photosynthetic light-harvesting pigment-protein complex B800-850 of Rhodopseudomonas capsulata, *Febs Lett.*, 182, 260, 1985.
32. **Wacker, T., Gadón, N. Becker, A., Mäntele, W., Kreutz, W., Drews, G., and Welte, W.**, Crystallization and spectroscopic investigation with polarized light of the reaction center-B875 light harvesting complex of *Rp. palustris*, *Febs Lett.*, 197, 267, 1986.

33. **Lee, J. C. and Timasheff, S. N.**, The stabilization of proteins by sucrose, *J. Biol. Chem.*, 256, 7193, 1981.
34. **Arakawa, T. and Timasheff, S. N.**, Stabilization of protein structure by sugars, *Biochemistry*, 21, 6536, 1982.
35. **Lehmann, M. S. and Zaccai, G.**, Neutron small-angle scattering studies of ribonuclease in mixed aqueous solutions and determination of the preferentially bound water, *Biochemistry*, 23, 1939, 1984.
36. **Saito-Nakatsuka, K., Yamashita, T., Kubota, I., and Kawakita, M.**, Reactive sulfhydryl groups of sarcoplasmic reticulum ATPase, *J. Biochem.*, 101, 365, 1987.
37. **Garavito, R. M., Markovic-Housley, Z., and Jenkins, J. A.**, The growth and characterization of membrane crystals, *J. Cryst. Growth*, 76, 701, 1986.
38. **Zulauf, M. and Rosenbusch, J. P.**, Micelle clusters of octylhydroxy oligo(oxyethylenes), *J. Phys. Chem.*, 87, 856, 1983.
39. **Zulauf, M., Weckström, K., Hayter, J. B., Degiorgio, V., and Corti, M.**, Neutron scattering study of micelle structure in isotropic aqueous solutions of poly(oxyethylene) amphiphiles, *J. Phys. Chem.*, 89, 3411, 1985.
40. **Karlström, G.**, A new model for upper and lower critical solution temperatures in poly(ethyleneoxide) solutions, *J. Phys. Chem.*, 89, 4962, 1985.
41. **Bartolino, R., Meuti, M., Chidichimo, G., and Ranieri, G. A.**, Lyotropic liquid crystals, in *Physics of Amphiphiles: Micelles, Vesicles and Microemulsions*, Degiorgio, V. and Corti, M., Eds., North-Holland, Amsterdam, 1985, 524.
42. **Weckström, K.**, Aqueous micellar systems in membrane protein crystallization, *Febs Lett.*, 192, 220, 1985.
43. **Parish, C. R., Classon, B. J., Tsagaratos, J., Walker, I. D., Kirszbaum, L., and McKenzie, I. F. C.**, Fractionation of detergent lysates of cells by ammonium sulfate-induced phase separation, *Anal. Biochem.*, 156, 495, 1986.
44. **Brooks, D. E., Sharp, K. A., and Fisher, D.**, Theoretical aspects of partitioning, in *Partitioning in Aqueous Two-Phase Systems*, Walter, H., Brooks, D. E., and Fisher, D., Eds., Academic Press, Orlando, FL, 1985, Chap. 2.
45. **Pittz, E. P. and Timasheff, S. N.**, Interaction of ribonuclease A with aqueous 2-methyl-2,4-pentanediol at pH 5.8, *Biochemistry*, 17, 615, 1978.
46. **Zeppezauer, M.**, Formation of large crystals, *Meth. Enzym.*, 20, 253, 1971.
47. **Weber, B. H. and Goodkin, P. E.**, A modified microdiffusion procedure for the growth of single protein crystals by concentration-gradient equilibrium dialysis, *Arch. Biochem. Biophys.*, 141, 489, 1970.
48. **McPherson, A., Koszelak, S., Axelrod, H., Day, J., Williams, R., Robinson, L., McGrath, M., and Cascio, D.**, An experiment regarding crystallization of soluble proteins in the presence of β-octyl glucoside, *J. Biol. Chem.*, 261, 1969, 1986.
49. **Garavito, R. M., and Rosenbusch, J. P.**, Three dimensional crystals of an integral membrane protein: an initial X-ray analysis, *J. Cell Biol.*, 86, 327, 1980.
50. **Albertsson, P., Cajarville, A., Brooks, D. E., and Tjerneld, F.**, Partition of proteins in aqueous polymer two-phase systems and the effect of molecular weight of the polymer, *Biochim. Biophys. Acta*, 926, 87, 1987.
51. **Arakawa, T. and Timasheff, S. N.**, Mechanism of polyethyleneglycol interaction with proteins, *Biochemistry*, 24, 6756, 1985.
52. **Albertsson, P.**, *Partition of Cell Particles and Macromolecules*, Almqvist & Wiksell, Stockholm, 1971.
53. **Wacker, T., Steck, K., Becker, A., Drews, G., Gadón, N., Kreutz, W., Mäntele, W., and Welte, W.**, Crystallization and spectroscopic investigations of the pigment-protein complexes of Rp. palustris, in Progress in Photosynthesis Research, Vol. I, Biggins, J., Ed., Martinus Nijhoff, Dordrecht, 1987, 383.
54. **Welte, W., Leonhard, M., Diedericks, K., Weltzien, H.-U., Restall, C., Hall, C., and Chapman, D.**, Stabilization of detergent-solubilized Ca^{2+} ATPase by poly(ethyleneglycol), *Biochem. Biophys. Acta.*, 984, 193, 1989.

Chapter 6

CRYSTALLIZATION OF PURPLE BACTERIAL ANTENNA COMPLEXES

Richard J. Cogdell, Kevin J. Woolley, Linda A. Ferguson, and Deborah J. Dawkins

TABLE OF CONTENTS

I. INTRODUCTORY REMARKS

We are currently involved in trying to understand the molecular details of the light-harvesting process in purple photosynthetic bacteria. It is quite clear that in order to achieve this aim we need to obtain a high resolution three-dimensional structure for a bacterial antenna complex. Unfortunately, it is still true that the only reliable way to determine such a high-resolution structure for a protein is to use the methods of X-ray crystallography. This of course requires the production of suitably sized, well ordered crystals and this is why we began to try and crystallize the bacterial antenna complexes.[1,2] In this chapter, after a brief introduction to bacterial antenna complexes, we shall describe some of our experiences in this area of crystallizing membrane proteins, with the hope that they will be of help to others who embark on a similar research program.

II. INTRODUCTION TO BACTERIAL ANTENNA COMPLEXES

The light-reactions of bacterial photosynthesis take place in and on the highly pigmented intracytoplasmic membranes.[3,4] The actual light-absorbing pigments (bacteriochlorophylls and carotenoids) are organized into two main types of pigment-protein complex.[5,6] The majority form the light-harvesting system and funnel absorbed radiant energy to a specialized few, called reaction centers, where that energy is trapped and used to drive photosynthesis. The bacteriochlorophylls and carotenoids are noncovalently bound to intrinsic membrane proteins, and whether a given pigment molecule is destined to fulfill a reaction center or antenna function is solely determined by which apoprotein it is bound to.

Purple photosynthetic bacteria, as a group, can produce two major classes of antenna complex.[7] The first type is usually present in a constant ratio with respect to the reaction center and forms the core of the photosynthetic unit. The second type, of which there are several different forms, is the variable portion of the photosynthetic unit, and its amount is controlled by the light-intensity at which the cells are grown. The fixed antenna type shows a single bacteriochlorophyll absorption band in the near infrared region (NIR) while the variable types show two strong NIR bacteriochlorophyll absorption bands. A great deal is known about the composition of these bacterial antenna complexes and this is summarized in Table 1 for a typical representative of each class. Also in many cases now the antenna apoproteins (the α- and β-apoproteins) have been sequenced.[8] The antenna apoproteins are small (50 to 65 amino acids long), very hydrophobic polypeptides, and they appear to have a single membrane spanning domain which is probably folding into an α-helix.[9,10]

The structure of the bacterial reaction center isolated from *Rhodopseudomonas viridis* could be determined to a resolution of 3 Å.[11,12] The success of that project has effectively dispelled the myth that it is impossible to crystallize membrane proteins and has really inspired other workers interested in the structure and function of membrane proteins to commit both time and effort to this area of membrane biochemistry. Unfortunately the reaction center is still the only case where a high-resolution structure of a membrane protein has been produced. It appears to be more typical that crystals of membrane proteins can be grown, but they tend not to be very well ordered.

The bacterial antenna complexes are rather ideal proteins to use to study the general methods for crystallizing membrane proteins. Many intrinsic membrane proteins are only present in the membrane in rather low concentrations. They are, therefore, often rather difficult to isolate and purify, moreover in many cases also lose their function when they are removed from the membrane. In contrast to this the bacterial antenna complexes are major membrane proteins and in some cases can represent up to 50% of the total membrane protein. The antenna complexes are also highly pigmented so that they can be easily followed during their purification by eye, thus removing the need for complicated assays. When the

TABLE 1

A Comparison of the B890-Complex From *Rhodopspirillum rubrum* S1 and
the B800-850-Complex From *Rhodobacter sphaeroides* 2.4.1

Antenna type	Bch*la*: Carotenoid ratio	Numbers of poly-peptides in isolated complex	Size of isolated polypeptides
B890-complex	2:1	1α	α-52 amino acids
		1β	β-54 amino acids
B800-850 complex	2:1	1α	α-54 amino acids
		1β	β-50 amino acids

pigments are correctly bound to their antenna apoproteins their absorption spectra are red-shifted with respect to their spectra in organic solvents. Any denaturation reverses this red-shift and can be easily seen by eye. This provides an excellent monitor for the integrity of the antenna complexes throughout the isolation, purification, and subsequent crystallization procedure. Finally, antenna complexes can now be isolated from quite a range of species of purple bacteria. This then allows a family of closely related proteins to be tested for ease of crystallization and greatly increases the chances of finding at least one that will be suitable.

III. ISOLATION OF THE B800-850-COMPLEXES

Before the process of crystallization of any protein can be undertaken that protein must be available in a highly purified form in milligram amounts. In this section we are only going to describe the purification of the B800-850-complex from *Rhodobacter sphaeroides* strain 2.4.1. There are differences in detail in the procedures used for the other complexes we study, but all the major points can be conveniently illustrated with references to *Rb. sphaeroides*.

We have found, purely by trial and error, that the detergent LDAO (lauryldimethylamine-*N*-oxide) is the most convenient to use (it is also rather inexpensive). The photosynthetic membranes are adjusted to the required concentration (OD_{850} of 50 cm^{-1}) and the reaction centers are selectively solubilized by treating the membranes with 0.25% v/v LDAO at 26°C for 30 minutes (overnight at 4°C will also work quite well).[13] A high speed centrifugation now pellets the membrane residue (containing the B800-850-complexes) and leaves the reaction centers in the supernatant. We normally save the reaction centers for other experiments. The pellets are now resuspended in buffer, readjusted to an OD_{850} of 50 cm^{-1} and solubilized with 1% v/v LDAO at room temperature. A low speed centrifugation then removes any unsolubilized material. The supernant from this spin is diluted with buffer to lower the LDAO concentration to 0.2% v/v and then is loaded on to a DEAE cellulose ion exchange column. The complex will only bind to the column if the pH of the Tris buffer is 8.0 or above, and if the detergent concentration is less than 0.5% v/v. The column is now developed in buffer containing 0.1% v/v LDAO with increasing concentrations of sodium chloride. The B800-850-complex usually elutes at between 150 to 250 m*M* sodium chloride, though the exact point of elution does vary significantly from preparation to preparation.

The fractions collected from the column are examined spectrophotometrically and the "good" ones pooled. Figure 1 shows a typical absorption spectrum of the B800-850-complex. We have annotated this spectrum to indicate the points to look for that we have been able to equate with a good quality preparation.

The pooled B800-850 fractions are now diluted with Tris buffer and reapplied to a second DEAE cellulose column. This is used to concentrate the complex prior to its further purification by molecular sieve chromatography. The concentrated complex is now passed down a 1 m long S-200 Sephacryl column. Again the fractions are examined spectrophotometrically

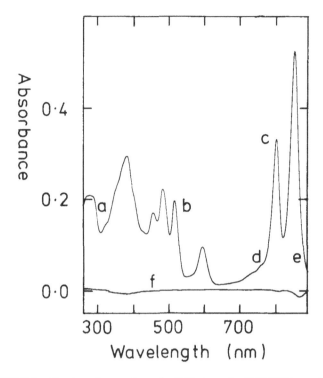

FIGURE 1. The absorption spectrum of a purified sample of B800-850-complex from *Rhodobacter sphaeroides*. The spectrum was recorded in a Pye-Unicam SP8-500 spectrophotometer and the sample was resuspended in 10 mm Tris HCl pH 8.0, 0.1% v/v LDAO. (a) Absorption here should be low compared to that at 850 nm. Best samples have a rather flat peak here. (b) This is where the carotenoids absorb and their peaks should be sharp and well-defined. (c) The 800 nm Bchl*a* absorption band is rather labile. Good samples have an 850 to 800 ratio of ~1.5 to 1.0. (d) There should be no 7750 nm absorption peak. That would be due to "free" Bchl*a*. (e) At 890 nm the absorption must be very much lower than the trough between the 800 and 850 nm peaks. This shows that B875 has been successfully removed. (f) Baseline.

and the good ones pooled. The pooled complex is now concentrated again on a small DEAE cellulose column. We usually use a small column poured in a pasteur pipette.

Figure 2 shows the polypeptide composition of the complex as it is taken through this purifications procedure. After the first ion exchange column the column is quite pure with only a few contaminating polypeptides present. These contaminants are then finally removed by the molecular sieving. By comparing the polypeptide composition of the complex during its isolation and purification we have found that a spectrophotometric measurement of the ratio of the absorbance at 850 nm (due to bacteriochlorophyll *a*) to that at 270 nm (due to aromatic amino acids) is a reliable, and easily measured indicator of purity. Ratios of 2:6:1 or bigger yield preparations that do crystallize and which are free of contaminating polypeptides on SDS polyacrylamide gels.

The reader should be aware that there are usually several alternative methods of isolating and purifying membrane proteins such as the bacterial antenna complexes, that will both work routinely and that can also be scaled up to produce milligram amounts. Do not be frightened to try several methods. We have recently also used an initial separation on a sucrose gradient centrifugation run to produce the B800-850-complex for the first ion exchange column, rather than the selective removal of the reaction centers.[14] This procedure also yields excellent preparations of the B800-850-complex.

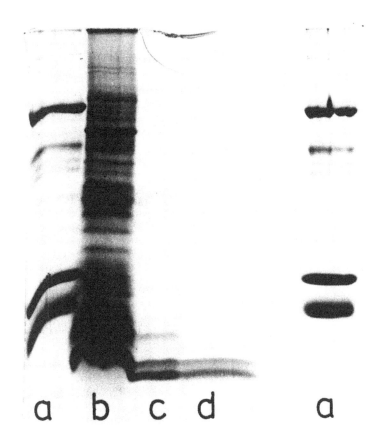

FIGURE 2. An SDS polyacrylamide gradient gel showing the state of purification of the B800-850-apoproteins from *Rhodobacter sphaeroides* during their isolation. This is an 11.5 → 16.5% polyacrylamide gradient gel and the samples have been stained with Coomassie blue. (a) Standard proteins, Bovine Serum Albumin (68KD); Alcohol Dehydrogenase (41KD); Myoglobin (17.2KD); Cytochrome *c* (11.7KD). (b) Solubilized membranes. (c) B800-850-fraction from the first DE52 column. (d) B800-850-complex after the molecular sieve column.

IV. CRYSTALLIZATION

A. INTRODUCTION

The basis of the method we use to crystallize the antenna complexes relies upon the addition of small amphiphiles.[15,16] We normally set up our crystallization attempts using either ammonium sulfate or potassium phosphate as the precipitants. In the absence of amphiphiles (such as heptane 1,2, 3 triol) addition of these salts induces a phase separation, with a detergent-rich phase and a major aqueous, detergent-poor phase. All the antenna complexes go into the detergent-rich phase and there they usually denature rather quickly. This process occurs at salt concentrations well below that required to cause true precipitation. Crystal formation tends to take place close to the point of precipitation and so under these conditions crystal formation is strongly inhibited. Addition of the amphiphile perturbs the phase diagram of the detergent so that phase separation now occurs at salt concentrations above those required to induce protein precipitation. Crystal formation is now possible.

Both Michel[14] and Garavito[15] have also emphasized a second role of the small amphiphiles in the a actual process of crystal formation. They view these small amphiphiles as agents which enter the crystal lattice and act to balance the hydrophilic and hydrophobic

TABLE 2

The Range of Antenna Complexes That We Currently Use in Our Crystallization Trials and a Brief Summary of How They Are Purified

Type of complex	Detergent used	Purification methods
B800-850 *Rb. sphaeroides*	LDAO	Column chromatography on DEAE-cellulose then molecular sieve chromatography
B800-850 *Rps. acidophila* 7750	LDAO	Column chromatography on DEAE-cellulose then molecular sieve in chromatography
B800-850 *Rps. gelatinosa*	LDAO	Sucrose gradient centrifugation and molecular sieve chromatography
B800-850 *Rps. blastica*	LDAO	Sucrose gradient centrifugation and molecular sieve chromatography
B800-820 *Rps. acidophila* 7050	LDAO	Sucrose gradient centrifugation and molecular sieve chromatography
RC-B890 conjugate *R. rubrum*	Lauryl-maltoside	Sucrose gradient centrifugation and molecular sieve chromatography
RC-B890 conjugate *Rps. gelatinosa*	LDAO	Sucrose gradient centrifugation and molecular sieve chromatography

surfaces of the membrane proteins. In this way these amphiphiles affect the packing of the proteins within the crystal. We (as described below) strongly support this view since we can obtain different crystal forms just by changing the type of amphiphile used.

B. CHOICE OF CONDITIONS FOR CRYSTALLIZATION TRIALS

Before sensible crystallization trials can be undertaken the conditions over which your protein is stable must be determined. We tested our antenna complexes for their stability with various salts, types of detergent and amphiphile, and with respect to pH and temperature. We quickly discovered that the most important parameters were pH and the choice of detergent. It is important to emphasize in this context that it is long term stability over a time period of weeks that is important, if large crystals are to be given a chance of developing. Using mainly salts as precipitating agents we have found that long-term stability requires the use of pHs above 8.5.

C. CHOICE OF DETERGENT

Most of our antenna preparations (see Table 2) initially involve the use of LDAO. This is a small switterionic detergent with a low critical micelle concentration (0.05 to 0.1% v/v) and which tends to form rather small micelles. In the initial stages of our preparations we use the inexpensive commercial LDAO obtained from the Onxy Chemical Corp., but for the crystallization trials we always use the purer Fluka product. One other good feature of this detergent is that it can readily be changed following binding of the antenna complex to a DEAE-cellulose ion exchange column. We usually do this on a mini-column poured in a Pasteur pipette. The antenna complex is applied to the column and binds. The column is then washed with detergent free buffer for about 30 minutes which reduces the amount of LDAO present. The last traces of LDAO are then removed chemically by washing the column with 1 column volume of buffer containing 0.5 mM sodium dithionite. This reductant reduces the amine-oxide bond in LDAO and removes the final traces of LDAO. The column is now washed with one column volume of buffer containing the required new detergent and the antenna complex is eluted by raising the salt cooncentration to 350 mM.

Using this method we have investigated effect of detergent type on both stability and crystal formation with the B800-850-complex from *Rhodopseudomonas acidophila* strain 7750. The results are presented in Table 3. So far we appear to get optimal crystal formation with either LDAO or DDAO (dimethyldecylamine-N-oxide). There is a lot of mythology

TABLE 3
A Summary of the Results of Trying to Crystallize the B800-850-Complex From *Rhodopseudomonas acidophila* Strain 7750 in the Presence of Different Detergents

Type of detergent	Nature of detergent	Crystal formation
LDAO	Zwitterionic	+ + +
DDAO	Zwitterionic	+ + + +
Lauryl-maltoside	Sugar-hydroxy	+
Cholic acid	Anionic	+
β-octyl-glucoside	Sugar-hydroxy	+ (slight denaturing effect)
n-octyl-rac-2,3-dioxypropylsulphoxide	−	+ +
Triton X-100	Non-ionic	− (we have never seen crystal formation with this detergent)
Sodium dodecylsulphate	Anionic	− (strongly denaturing)

Note: (−): no crystal formation. (+): crystal formation, the larger the number of crosses the better the crystal formation.

that has built up about which detergents will or will not support crystal formation of membrane proteins. In our hands the bacterial antenna complexes crystallize from most of the detergents that we have tested, so long as they are stable in them. In general though we find the zwitterionic ones to be the most useful. The only one we have tried and consistently failed with is Triton X-100, even though the complexes are very stable in this detergent. Our advice in this area is to try every detergent that you can lay your hands on and not to be overly influenced by what has been found to work in other systems.

D. CHOICE OF SMALL AMPHIPHILE

Our first success in obtaining crystals with a bacterial antenna complex was using the B800-850-complex from *Rps. acidophila* strain 7750. The complex was made up with LDAO, ammonium sulfate as the precipitant and with 3% v/v heptane-triol (the high melting point isomer) as the small amphiphile. As the sample was concentrated by vapor diffusion against a more concentrated solution of ammonium sulfate thin, needle-like crystals were obtained (Figure 3). We initially tried six different types of complex and one, the B800-850-complex from *Rps. acidophila*, actually worked immediately.

We have subsequently tried a range of other small molecules to see if they would act as effective small amphiphiles. Our conclusions from these studies are shown in Table 4. So far benzamidine hydrochloride (better known as a protease inhibitor) has proved to be the most useful. Figure 4 shows some of the crystal forms that we have obtained using the various amphiphiles with the B800-850-complex from *Rps. acidophila*.

E. CHOICE OF PRECIPITANT

So far we have tested ammonium sulfate, potassium phosphate, and various polyethylene glycols (PEG) for their suitability as precipitants to induce crystal formation. With the B800-850-complexes we have only been successful in obtaining crystals with the salts as precipitants not the PEGs. However, we have found PEG 2000 a useful precipitant with the B880-RC conjugates.[17]

Figure 5 also shows some of the crystals that we have obtained with the different precipitants. It is worth noting again that we have only been successful with ammonium sulfate at pHs above 8.5.

F. CHOICE OF TEMPERATURE

We have investigated the effects of growing our crystals at 4, 10, and 20°C. At 20°C the crystals do grow faster, as would be expected from the increase in the rate of vapor

FIGURE 3. Crystals of the B800-850-complex from *Rhodopseudomonas acidophila* strain 7750 grown in the presence of heptane-triol with ammonium sulfate as the precipitant. The bar represents 100 μ. The antenna sample was set up in 2 *M* ammonium sulfate, pH 9.5 with 0.1% v/v LDAO, 2.5% heptane triol and 20 m*M* Tris HCl. This sample was allowed to concentrate by vapor diffusion at 4°C against 2.8 *M* ammonium sulfate which was also adjusted to pH 9.5.

TABLE 4
A Summary of the Effectiveness of Some of the Molecules That We Have Tried to Use as Small Amphiphiles With the B800-850-Complex From *Rhodopseudomonas acidophila* Strain 7750 to Promote Crystal Formation

Amphiphiles	Precipitant	Type of crystals formed
Heptane 1,2,3-triol	Ammonium sulphate	Long, thin needles
Heptane 1,2,3-triol + piperidine-2-carboxylic acid	Potassium phosphate	Short, chunky needles
Piperidine-2-carboxylic acid	Potassium phosphate	Flat, square plates
Benzamidine-hydrochloride	Potassium phosphate	1. Flat plates[a] 2. Tetragonal 3. Chunky needles
Tryptopan	Potassium phosphate	Thin needles
L-Proline	Potassium phosphate	Very small needles
L-Threonine	Potassium phosphate	Small square plates
Octane 1,2,3-triol	Potassium phosphate	Thin needles
Hexane 1,2,3-triol	Potassium phosphate	No crystals obtained

[a] With benzamidine hydrochloride we get different crystal forms depending upon both the temperature and pH at which we grow them.

diffusion. However we have reached no definite conclusions as yet as to which temperature is best and so we always set up identical trials at each temperature. We have had some indication in the case of the B800-850-complex from *Rb. sphaeroides* strain 2.4.1 that at 20°C we get a different crystal form than at 4°C, but this finding is somewhat variable at present (see Figure 5).

G. EFFECT OF PROTEIN CONCENTRATION
We normally set up our crystallizations at protein concentrations between 2 and 5 mg/

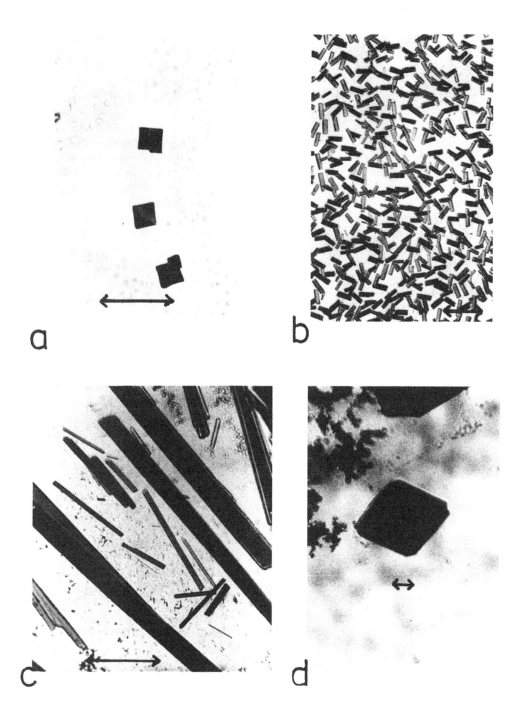

FIGURE 4. The effect of changing the amphiphile upon the crystal form of the B800-850-complex of *Rhodo-pseudomonas acidophila* strain 7750. In each case the bar represents 100 μ. These crystals were grown in the presence of phosphate as the precipitant and LDAO as the detergent. (a) With 3% piperidine-2-carboxylic acid as the amphiphile. (b) With 2% heptane-triol and 3.5% piperidine-2-carboxylic acid as the amphiphiles. (c) With 3.5% benzamidine hydrochloride as the amphiphile at pH 9.5. (d) With 3.5% benzamidine hydrochloride as the amphiphile at pH 10.5.

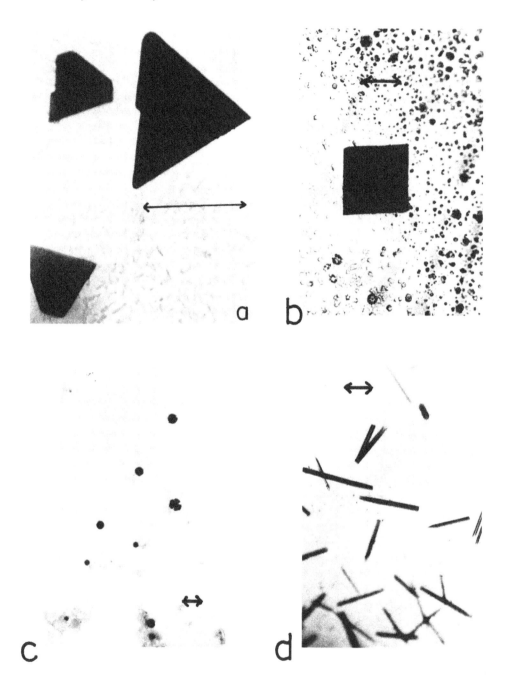

FIGURE 5. The use of different precipitants to crystallize bacterial antenna complexes. In each case the bar represents 100 μ. (a) Crystals of the RC-B880 conjugate from *Rhodospirillum rubrum*, grown with PEG 2000 as the precipitant. (b) Crystals of the B800-850-complex from *Rb. sphaeroides* grown with phosphate as the precipitant at 4°C. (c) Crystals of the B800-850-complex from *Rb. sphaeroides* grown with phosphate as the precipitant at 20°C. (d) Crystals of the B800-850-complex from *Rps. acidophila* strain 7750 grown with ammonium sulfate as the precipitant.

ml. Much lower than 1 mg/ml and the crystals run out of protein very quickly and remain very small, much more than 10 mg/ml we tend to get a large amount of amorphous precipitation occurring. There have been reports[18] where different crystal forms have been obtained just by changing the protein concentration. We have not yet seen this under our crystallization conditions.

V. EXPERIMENTS WITH THE CRYSTALS

So far we have been able to obtain single crystals from different complexes which are large enough to be examined in an X-ray beam. Very recently one crystal form of the B800-850-complex from *Rps. acidophila* strain 10050 > was obtained diffracting X-rays to 3.5 Å resolution.[19] β-Octylglucoside was used as detergent and benzamidine as small amphiphile, potassium phosphate as precipitant.

Even in the absence of well-ordered crystals (i.e., to crystals that diffract to beyond 4 Å) there are still many indirect studies that can be carried out on thin crystals. Some of our crystal forms are optically very dichroic. This is illustrated in Plate 2* where for two different crystals we have photographed them under polarized light. Chapter 5 in this book should be consulted for further details on this aspect of the topic.

REFERENCES

1. **Cogdell, R. J., Woolley, K. J., Mackenzie, R. C., Lindsay, J. G., Michel, H., Dobler, J., and Zinth, W.,** Crystallization of the B800-850-complex from *Rhodopseudomonas acidophila* strain 7750′ in Springer Series in *Chem. Phys.*, 42, 85, 1985.
2. **Woolley, K. J., Mackenzie, R. C., Cogdell, R. J., Lindsay, J. G., and Michel, H.,** Crystallization of the B800-850-complex from *Rhodopseudomonas acidophila* strain 7750, *Biochem. Soc. Trans.*, 14, 57, 1986.
3. **Remsen, C. C.,** Comparative subcellular architecture of photosynthetic bacteries in *The Photosynthetic Bacteria*, Clayton, R. K. and Sistrom, W. R., Eds., Plenum Press, New York and London, 1978, 31.
4. **Drews, G.,** Structure and functional organisation of light-harvesting complexes and photochemical reaction centres in membranes of phototrophic bacteria, *Microbiol. Rev.*, 49, 59, 1985.
5. **Cogdell, R. J. and Thornber, J. P.,** Light-harvesting pigment-protein complexes of purple photosynthetic bacteria, *FEBS Lett.*, 122, 1, 1980.
6. **Thornber, J. P., Cogdell, R. J., Pierson, B. K., and Seftor, R. E. B.,** Pigment-protein complexes of purple photosynthetic bacteria: an overview, *J. Cell. Biochem.*, 23, 159, 1983.
7. **Cogdell, R. J.,** Light-harvesting complexes in the purple photosynthetic bacteria in *Encyclopaedia of Plant Physiology New Series*, Vol. 19, Staehelin, L. A. and Arntzen, C. J., Eds., Springer-Verlag, Berlin, Heidelberg, 1986, 252.
8. **Zuber, H.,** Primary structure and function of the light-harvesting polypeptides from cyanobacteria, Red Algae and Purple Photosynthetic Bacteria in *Encyclopaedia of Plant Physiology New Series*, Vol. 19, Staehehin, L. A. and Arntzen, C. J., Eds., Springer-Verlag, Berlin, Heidelberg, 1986, 238.
9. **Breton, J. and Nabedryk, E.,** Transmembrane orientation of α-helices and the organization of chlorophylls in photosynthetic pigment-protein complexes, *FEBS Lett.*, 176, 355, 1984.
10. **Cogdell, R. J. and Scheer, H.,** Circular dichroism of light-harvesting complexes from purple photosynthetic bacteria, *Photochem. Photobiol.*, 41, 1, 1985.
11. **Deisenhofer, J., Eppo, O., Miki, K., Huber, R., and Michel, H.,** X-ray structure analysis of a membrane protein complex. Electron density map at 3 Å resolution and a model of the chromophores of the photosynthetic reaction centre of *Rhodopseudomonas viridis*, *J. Mol. Biol.*, 180, 385, 1984.

* Plate 2 appears after page 136.

12. **Deisenhofer, J., Epp, O., Miki, K., Huber, R., and Michel, H.,** Structure of the protein subunits in the photosynthetic reaction centre of *Rhodopseudomonas viridis* at 3 Å resolution, *Nature,* 318, 618, 1985.

13. **Jolchine, G. and Reiss-Husson, F.,** Comparative studies on two reaction centre preparations from *Rps. sphaeroides* Y., *FEBS Lett.,* 40, 1974.

14. **Firsow, W. N. and Drews, G.,** Differentiation of the intracytoplasmic membrane of *Rps. palustris* induced by variation of oxygen partial pressure or light-intensity, *Arch. Microbiol.,* 115, 299, 1977.

15. **Michel, H. and Oestorhelt, D.,** Three dimensional crystals of membrane proteins: Bacteriorhodopsin, *Proc. Natl. Acad. Sci. USA,* 77, 1283, 1980.

16. **Garavito, R. M. and Rosenbusch, J. P.,** Three-dimensional crystals of an integral membrane protein: an initial X-ray analysis, *J. Cell. Biol.,* 86, 327, 1980.

17. **Dawkins, D. J., Ferguson, L. A., and Cogdell, R. J.,** Isolation and characterization of reaction centre-antenna conjugates from a range of bacteriochlorophyll *a*-containing species of purple bacteria, in *Microbial Energy Transduction,* Current Communications in Molecular Biology, Cold Springs Harbor Laboratory, 1986, 47.

18. **Allen, J. P., Theilder, R., and Feher, G.,** Crystallization and linear dichroism measurements of the B800-850 antenna pigment-protein complex from *Rhodopseudomonas sphaeroides* 2.4.1., in Springer Series in *Chem. Phys.,* 42, 82, 1985.

19. **Papiz, M. Z., Hawthronthwaite, A. M., Codgell, R. J., Wooley, K. J., Wightman, P. A., Ferguson, L. A., and Lindsay, J. G.,** Crystallization and characterization of two crystal forms of the B800-850 light-harvesting complex from *Rhodopseudomonas* acidophila strain 10050, *J. Mol. Biol.,* 209, 833, 1989.

Plate 2. Crystals of the B800-850 complex from *Rhodopseudomonas acidophila* strain 7750 photographed under polarized light. These crystals were grown in the presence of benzamidine hydrochloride with phosphate as the precipitant. The two frames were photographed with the direction of polarization rotated by 90°. The bar represents 100 μ.

Chapter 7

CRYSTALLIZATION OF REACTION CENTERS FROM *RHODOBACTER SPHAEROIDES*

James Paul Allen and George Feher

TABLE OF CONTENTS

I. INTRODUCTION

The conversion of light into chemical energy is mediated in photosynthetic organisms by a pigment-protein complex called the reaction center (RC). The existence of such an entity was first postulated by Emerson and Arnold in 1932[1] although direct evidence for their presence was not forthcoming until 1952.[2] Since this early work, a great deal of progress has been made on the isolation, purification, and characterization of the RC unit.[3-5] In particular, the RC from *Rhodobacter sphaeroides* has been well characterized and techniques for manipulating it have been developed to a high degree (e.g., cofactors, including the iron, can be removed and replaced, subunits can be dissociated, etc.) (for a review see Reference 6).

Despite this progress, the lack of knowledge of the exact spatial relationship between the reactants had made it difficult to understand quantitatively the details of the energy transfer process, in particular the remarkably high quantum efficiency of the photosynthetic process. To determine the three dimensional structure by X-ray diffraction, it is necessary to obtain well-ordered single crystals of the protein. However, until recently, many believed that integral membrane proteins could not be crystallized due to the presence of randomly oriented detergent molecules needed to solubilize the protein. It was argued that the lack of identical protein-detergent complexes would prevent the formation of an ordered crystalline array. This concept was proven to be wrong by the initial reports of low resolution diffraction from single crystals of bacteriorhodopsin from *Holobacterium halobium*,[7] porin from *Escherichia coli*,[8] and the subsequent observations of diffraction at atomic resolution (at least 3.0 Å) from crystals of RCs from *Rhodopseudomonas viridis*[9] and porin from *E. coli*.[10] Soon thereafter, different crystal forms of X-ray diffraction quality were reported for RCs from *Rb. sphaeroides*.[11-16] Analysis of the diffraction data lead to the determination of the three dimensional structure of the RC from *Rps. viridis*.[17-19] It is the first membrane protein whose structure was determined at atomic resolution. Subsequently, the structures of the RCs from *Rb. sphaeroides* and *Rps. viridis* were shown to be homologous by the molecular replacement technique.[20-22] This was soon followed by the determination of the three dimensional structure of the RC from *Rb. sphaeroides* strain R-26,[23-30] and *Rb. sphaeroides* strain 2.4.1.[26,28] These structures have provided the framework needed to understand the electron transfer process as well as serving as models for the structures of other membrane proteins (for a review see Reference 31).

The crystallization of proteins is usually accomplished by a trial and error approach that is guided by a large body of empirically determined procedures and conditions.[32] Although a systematic approach to investigate the mechanisms of crystallization of water soluble proteins has been developed,[33] these investigations have so far not been extended to membrane proteins. In this article we report on our empirical results on the crystallization of the RC from *Rb. sphaeroides* strain R-26.* The conditions used to crystallize the RC in several different forms are compared and the sensitivity of the crystallization to various parameters is discussed. Emphasis is placed on the role of the detergents and amphiphiles during crystallization of membrane proteins, in particular, the detergent interactions that lead to a phase separation and that affect the ordering of the RC in the crystal.

II. METHODS

A. MATERIALS

RCs from *Rb. sphaeroides* were purified as previously described, using either the de-

---

* RCs purified from the wild type strain 2.4.1 as well as biochemically altered RCs (e.g., by the removal of a quinone) have been found to crystallize under the same conditions as the native RCs from the R-26 strain.

tergent N,N-dimethyldodecylamine-N-oxide (LDAO)[6] or β-octyl glucopyranoside (βOG).[34] LDAO was obtained from Onyx (Los Angeles, CA), βOG from Calbiochem (San Diego, CA), heptane triol, hexane triol, octane triol, and (undecyl, decyl, nonyl, octyl, heptyl, and hexyl) dimethylamine-N-oxides from Oxyl (Bobingen, BRG), triethylammonium phosphate (TEAP) from Pfaltz and Bauer (Stanford, CT), glycolate butyl ester (GBE) from M. Garavito, (University of Chicago, Chicago, Illinois), β-hexyl glucopyranoside (βHG) from Sigma (St. Louis, MO), and polyethylene glycol (PEG) 4000 from Wilshire Chemical (Gardena, CA).

B. CRYSTALLIZATION

All crystals were grown by the "sitting drop" technique in which the protein solution is equilibrated by vapor diffusion with a reservoir solution. All protein and reservoir solutions contained 1 mM EDTA and 0.1% NaN$_3$ in addition to varying amounts of detergent, buffer, and amphiphiles that are detailed in Section III. After centrifugation (8000 g for 2 minutes), the supernatant of the protein solution was pipetted (20 to 200 μl) into one of four crystallization wells. These wells were cut from an array of 48 wells that formed a tissue culture plate (Linbro Cat. No. 76-336-05, Flow Laboratories, Virginia). The reservoir solution (5 ml) consisted of a high concentration of salt. The two solutions were enclosed in a plastic box (4.5 × 4.5 × 2 cm³). The solvent was gradually transferred by vapor diffusion from the RC solution to the reservoir solution until the two were in equilibrium. For solutions containing both PEG and NaCl this equilibration was driven principally by the difference in NaCl concentrations (0.2 M NaCl equilibrates with ~20% PEG by vapor diffusion).

Most of the crystallizations were done at ambient (~22°C) temperature. Crystallization at lower temperatures (4 to 22°C) was accomplished by placing the crystallizing solutions in locally designed styrofoam boxes that contained thermoelectric modules (Midland Ross, Cambridge, MA); this design avoided the use of fans thereby eliminating potentially harmful vibrations.[33]

C. DIALYSIS OF CRYSTALS

To exchange the detergent present in the crystals, microdialysis cells similar to those introduced by Zeppenzauer[35] were used. A piece of dialysis membrane was placed at one end of a quartz tube and held by sliding the membrane and quartz tubing into a teflon collar (see Figure 1). The teflon collar also provided support legs for the system. The dialysis cell was placed into a glass beaker containing 600 μl of reservoir solution. Crystals and mother liquor (~40 μl) were inserted through the open end of the tubing and the crystals were placed onto the membrane. The beaker was sealed with a greased glass cover slide. Dialysis proceeded without stirring for 2 to 4 d with 1 to 2 exchanges of the reservoir solution.

III. RESULTS AND DISCUSSION

A. CRYSTALLIZATION CONDITIONS

Two sets of conditions were originally used to grow different crystal forms (A1-2, B1-3) (see Figures 2 and 3) of RCs from *Rb. sphaeroides*.[11] One set (using ammonium sulfate and LDAO) were similar to the conditions used to crystallize RCs from *Rps. viridis*.[9] The other set (using PEG, NaCl, and βOG) corresponded approximately to the conditions used to crystallize porin from *E. coli*.[8,10] These two sets of conditions were then empirically altered as detailed below. This resulted in the growth of several new forms (A3-5, B4-11). Form B6 was used to determine the three dimensional structure by X-ray diffraction.

1. Crystals Grown with Ammonium Sulfate

RCs were purified using the detergent LDAO. The starting conditions for the protein solution were: RC concentration 5.5 mg/ml, 0.1% LDAO, 2.5% heptane triol, 3% TEAP,

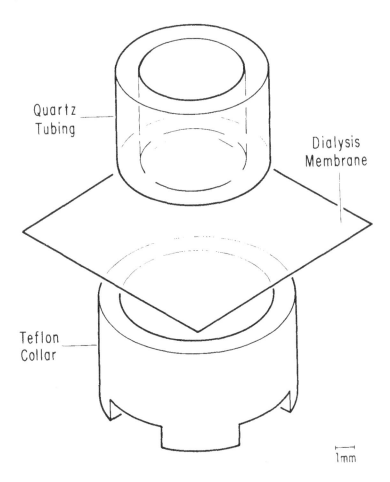

FIGURE 1.   Exploded view of the dialysis cell used for exchanging detergents in RC crystals. The quartz tube slides into the teflon collar with the dialysis membrane interposed.

and 1.2 M (NH$_4$)$_2$SO$_4$. After mixing the pH was adjusted to 7.0 with NaOH. The salt reservoir contained 2.1 M (NH$_4$)$_2$SO$_4$. Two crystal forms, A1 and A2, were observed (see Figure 2) after 2 to 4 weeks at room temperature.

Three different crystal forms were grown by independently changing two of the crystallization parameters described above: (1) When the concentration of the salt reservoir was lowered from 2.1 to 1.8 M (NH$_4$)$_2$SO$_4$, two new forms, A3 and A4, were observed (Figure 2); (2) when the concentration of heptane triol was lowered from 2.5 to 1.0%, a new diamond shaped form, A5, was observed (Figure 2).

2. Crystals Grown with Polyethylene Glycol and NaCl

RCs were purified using the detergent βOG. The starting conditions for the protein solution were: RC concentration 3.3 mg/ml, 8% PEG 4000, 0.8% βOG, 0.3 M NaCl, and 15 mM Tris-Cl pH 8.0. The salt reservoir consisted of 25% PEG 4000, 0.8 M NaCl, and 15 mM Tris-Cl pH 8.0. After 2 to 4 weeks at room temperature, three crystal forms, B1, B2, and B3, were observed (see Figure 3). A distinct phase boundary was usually observed after 4 d (see for example B1 of Figure 3).

Retaining βOG as the detergent, four different crystal forms were observed when the following changes of the original conditions were made: (1) when the amphiphile glycolate butyl ester (GBE) was added (4%), both small needles and thin aggregated rectangular plates were observed (forms B9 and B10 not shown); and (2) when the amphiphile heptane triol was added (3%), two new forms, B4 and B5, were observed (see Figure 3).

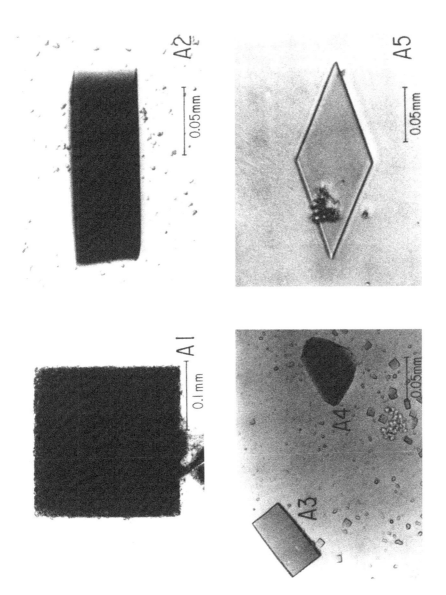

FIGURE 2. Different crystal forms of *Rhodobacter sphaeroides* grown in ammonium sulfate. For forms A1 and A2, the conditions of the protein solution were: RC concentration 5.5 mg/ml, 0.1% LDAO, 2.5% heptane triol, 3% TEAP, 1.2 M $(NH_4)_2SO_4$, pH 7.0; the reservoir solution contained 2.1 M $(NH_4)_2SO_4$. Forms A3 and A4 were observed when the concentraion of the salt reservoir was lowered from 2.1 to 1.8 M $(NH_4)_2SO_4$. Form A5 grew when the concentration of heptane triol was lowered from 2.5 to 1.0%.

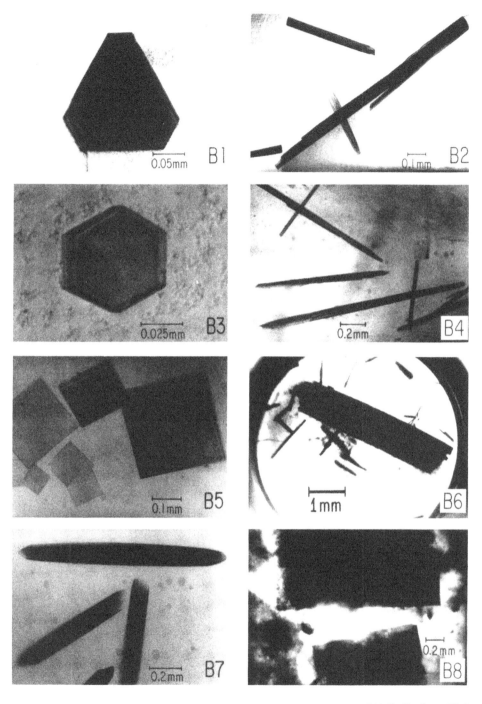

FIGURE 3.   Different crystal forms of *Rhodobacter sphaeroides* grown in PEG and NaCl. For forms B1-3, the conditions of the protein solution were: RC concentration 3.3 mg/ml, 8% PEG 4000, 0.8% βOG, 0.3 *M* NaCl, and 15 m*M* Tris-Cl pH 8.0; the reservoir solution contained 25% PEG, 0.8 *M* NaCl, and 15 m*M* Tris-Cl pH 8.0. Forms B4 and B5 grew when 3% heptane triol was added. Form B6 grew when the 0.8% βOG was replaced with 0.06% LDAO and 3.9% heptane triol was added. Form B7 grew when the 0.8% βOG was replaced with 0.06% LDAO, 3.9% heptane triol was added, and the temperature was lowered from 22 to 4°C. Form B8 grew when the 0.8% βOG was replaced with 2.5% NDAO and 4% heptane triol was added. Not shown are Forms B9 and B10 (grown when 4% GBE is added) or Form B11 (grown when 0.8% βOG is replaced by 0.06% LDAO, 3.9% heptane triol is added, and 3% TEAP or 3% βHG is added). B6 was the crystal form used to obtain the three dimensional structure of RCs from *Rb. sphaeroides*.[22-26]

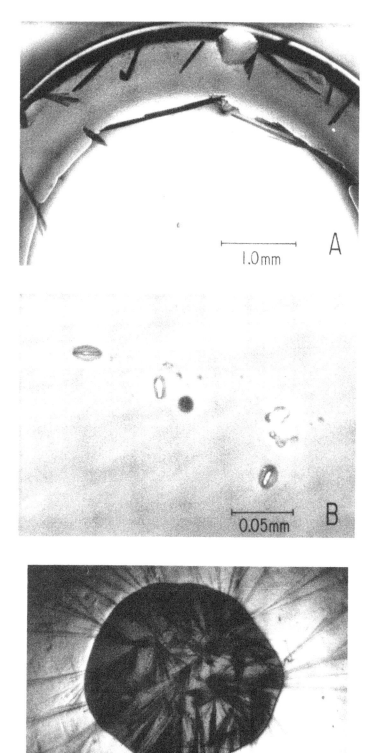

FIGURE 4. Three different types of phase separations seen in crystallization solutions containing PEG, NaCl, and the detergent βOG. Conditions as described in Section III.A.3.

Replacing the detergent βOG with other detergents lead to the growth of new forms, but only in the presence of heptane triol. When RCs were purified with the detergent LDAO and the crystallization conditions were changed from 0.8% βOG to 0.06% LDAO with 3.9% heptane triol crystal form B6 was obtained (see Figure 3). Modification of the LDAO and heptane triol solution lead to the growth of three new forms. (1) Adding either 3% TEAP or 3% β-hexyl glucoside lead to the growth of small squares (form B11, not shown, similar in shape to B5 and B8 but much smaller). (2) After lowering the temperature from 22 to 4°C, a new form, B7, was observed (see Figure 3). The cross section of these crystals viewed down the long axis is identical to that seen in form B1 of Figure 3. (3) The detergent LDAO was replaced with the detergent NDAO, whose carbon chain is C_9 rather than C_{12}. RCs in LDAO were dialyzed against 0.5% NDAO; the dialyzed RCs were added to the crystallization solution in which the concentration of NDAO was 2.5%. This lead to the growth of large squares; form B8 (see Figure 3) amid the presence of a large amount of amorphous precipitates.

3. Phase Separation

The hydrophobic nature of the carbon chain of detergents leads to aggregation in polar solvents under conditions that depend upon the specific detergent and solvent. At very low concentrations, the detergent molecules exist as monomers. When the detergent concentration is increased above a certain point, called the critical micelle concentration, the detergents aggregate into micelles in which the carbon chains are in contact with each other and the head groups are exposed to the polar solvent.[36] For βOG in water, the critical micelle concentration is ~0.8% w/v.[36] In the crystallization solution, the detergent βOG is present with the polymer PEG; for this solution an additional aggregation mechanism occurs. At the initial concentrations of 0.8% βOG and 8.0% PEG, the solutions are homogeneous (i.e., no phase separation). During crystallization both the βOG and PEG concentrations increase by a factor of ~3. This leads to a clustering of detergent micelles and the subsequent phase separation of the βOG region separated from the PEG region.[8,10,37] In the crystallization solutions, the RC is concentrated in the detergent region.

Three different types of phase separation were observed in the crystallization solutions of *Rb. sphaeroides* (see Figure 4). (1) The RC and detergent region formed along the walls of the well (see Frame A of Figure 4) for the conditions used to grow Forms B1-3 (protein solution containing 0.3 M NaCl, 0.8% βOG, 8.0% PEG, 3.3 mg/ml RC, and reservoir containing 0.8 M NaCl and 25% PEG). This type of phase separation was the most useful in growing large crystals. Some crystals grew in the outer region though many grew at the boundary between the two regions. The demarcation of the boundary was not always pronounced; however, a sharp boundary was often found for conditions that optimized the growth of crystals. (2) The RC and detergent collected into many small drops; crystals grew inside the drops (see Frame B of Figure 4). This behavior often occurred when the amphiphile, GBE was included in the crystallization solution. The formation of this type of phase separation was also evident when LDAO replaced the βOG without the addition of heptane triol, though no crystals were observed for these conditions. (3) An RC and detergent region formed in the center of the well; many thin crystals grew within this region (see Frame C of Figure 4). This was observed when the reservoir solution had a higher PEG concentration but a lower NaCl concentration than the protein solution. The conditions required for this type of phase separation were usually not favorable for crystallization.

The addition of an amphiphile to the crystallization solution often eliminated the phase separation. For instance, the phase separation obtained in solutions containing either βOG or LDAO with PEG (see Frame A of Figure 4) was less pronounced for solutions containing heptane triol, with no phase separation observed for concentrations of heptane triol of 4% and above. Thus, no phase separation was seen for solutions used to grow Forms B4, B5,

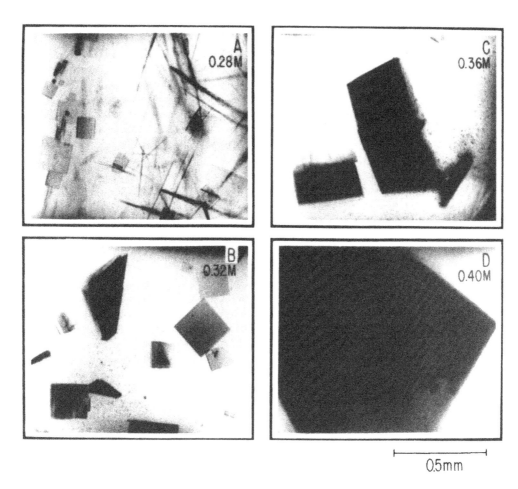

0.5mm

FIGURE 5. Effect of NaCl concentration on the crystallization of RCs (Forms B4 and B5). NaCl concentrations given in upper right corner. All other parameters kept constant as given in Figure 3.

B6, or B7. Similarly, no phase separation was seen in the NDAO and heptane triol solutions used to crystallize form B8.

B. SENSITIVITY OF THE CRYSTALLIZATION TO CHANGES IN PARAMETERS

The growth of the different forms was influenced by "traditional" crystallization parameters such as salt and protein concentrations, pH, and temperature. Crystallization of the different forms was observed within a certain range of these parameters about the original conditions, i.e., ± 300 mM for the salt concentration, ± 3.0 pH units, ± 8°C. For optimal growth, the RC concentration was often near the solubility limit; lower concentrations of the RC usually lead to the growth of smaller crystals. As a general rule, crystallization was most sensitive to a particular parameter when changes in that parameter caused a change in the crystal form. For example, both the number and size of forms B4 and B5 (see Figure 3) were very sensitive to the salt concentration of the protein solution (see Figure 5). At 0.28 M NaCl, many small crystals of form B4 (needles) and B5 (squares) were observed. Increasing the concentration of NaCl suppressed the growth of B4 and favored the growth of B5. At 0.4 M NaCl, only large crystals of form B5 (~1 mm × 1 mm × 0.2 mm) were observed.

The "nontraditional" parameters, those related to the detergents and amphiphile, also

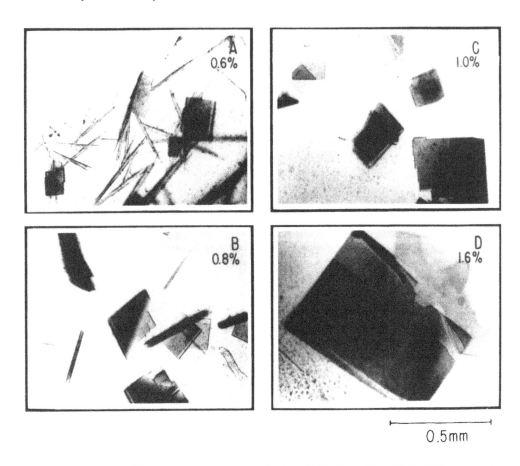

0.5mm

FIGURE 6.   Effect of βOG concentration on the crystallization of RCs (Forms B4 and B5). βOG concentration given in upper right corner. All other parameters kept constant as given in Figure 3.

influenced the growth of the different forms. The most striking effect was the change of form due to a change of detergent or addition of amphiphile as described in the next section. In some cases, the crystallization was also very sensitive to the detergent concentration (see Figure 6). For example, at 0.6% βOG, many needles (form B4) and some squares (form B5) were observed. Increasing the concentration of βOG suppressed the growth of form B4 and favored the growth of B5. At 1.6% βOG, only large crystals of form B5 were observed. Comparing Figures 5 and 6 shows that the effects due to changing the concentrations of βOG and NaCl are comparable. In other cases, crystallization was relatively insensitive to the detergent and amphiphile concentration. For instance, form B6 was grown with LDAO concentrations ranging from 0.02 to 0.50% and heptane triol concentrations from 2 to 7% (optimal conditions were 0.06% LDAO and 3.9% heptane triol).

   Another "nontraditional" parameter is the chain length of the detergent and amphiphile. This property strongly effects the formation of micelles[36] and the solubility of amphiphiles (for instance octane triol is less soluble than heptane triol in water).[38] In some cases, changing this parameter had little effect on the crystallization of the RC. For example, no significant change in crystal quality was observed when hexane triol was substituted for heptane triol or nonyl glucoside for octyl glucoside. In contrast, large effects were seen when the chain length of LDAO (C_{12} chain) was changed. Using the detergent undecyldimethylamine oxide (C_{11} chain) lead to minor changes in crystal growth of form B6. Using nonyldimethylamine oxide (C_9 chain) lead to the growth of new crystal form B8 (Figure 3). This new form was also grown in mixed detergent systems (LDAO and heptyldimethylamine oxide, C_7 chain).

C. COMPARISON OF DIFFERENT CRYSTALLIZATION CONDITIONS AND SPACE GROUPS

1. Effect of Crystallization Conditions on Crystal Form

The RC crystallized in a variety of forms as described above. The growth of different forms was due to changes in both the "traditional" and "nontraditional" crystallization parameters. The relationship between the crystallization conditions for each form is summarized in Figure 7. First, two independent sets of conditions were found that produced crystals (A1-2 and B1-3). Changes in a traditional parameter (arrow to left), the salt concentration, lead to two new forms (A3-4). Changes in other traditional parameters, pH and protein concentration, did not result in any form changes. Changes in nontraditional components, i.e., detergent (arrow to right) lead to the growth of forms A5, B4, B5, B9, and B10. Two combined changes (replacing βOG with LDAO and adding heptane triol) produced form B6. Several new forms (B7, B8, B11) were observed when the conditions for the optimized growth of form B6 were altered by either traditional (temperature) and nontraditional (detergent, amphiphile) parameter changes. Once each form was observed, all parameters were adjusted to achieve optimal growth.

2. Space Groups

Several of the different crystal forms grew to a sufficient size to identify their space groups by X-ray diffraction; the results are summarized in Table 1. The different crystals forms have approximately the same density V_m. Using a molecular weight of 150,000 Da for the RC and bound detergent, V_m varies from 2.1 to 2.7 Å^3/dalton; this agrees well with the most common value of 2.15 Å^3/dalton found for water soluble proteins.[39] Forms B5 and B6 both have the orthorhombic space group $P2_12_12_1$ with two cell dimensions, a and c, being nearly identical. The principal difference between these two space groups is the fourfold increase of the b axis of form B5 which reflects the fourfold increase of the asymmetric unit. Similarly, form B8 has an orthorhombic or tetragonal space group with cell dimensions of 140 × 140 × 280 Å^3; the precise space group has not been determined. Forms B2 and B4 both have the monoclinic space group P2. The fourfold increase in the asymmetric unit is mostly accommodated by the nearly twofold increase in both a and c. The remaining form A1 has the orthorhombic space group C222; this was the only one of these six that was grown in ammonium sulfate. The positions of the RCs in the unit cell has been characterized only for the form B6 (as discussed in the following section).

D. IMPROVEMENT OF CRYSTALLINE ORDER BY EXCHANGING DETERGENT

Some of the crystal forms, despite well formed facets and edges, exhibited a disorder in their X-ray diffraction patterns. In particular, Form B6 often exhibited a type of disordered diffraction pattern that may be indexed on a C-centered lattice with a doubling of the a and b axes. These additional reflections are not as sharp as the main reflections. This behavior indicates the presence of short range orientational disorder in the crystal packing.[40,41] A schematic representation of this type of disorder is shown in Figure 8. In a crystal with unit cell length a, every molecule is identically aligned (Case A in Figure 8). If every other molecule (the shaded molecules) is rotated (relative to the unshaded molecules) (Case B), the unit cell dimension of 2a results in a doubling of the number of diffraction spots at a given resolution as observed in the diffraction pattern. In case C, the crystalline array contains a disorder as some molecules at random positions are rotated. The probabilistic nature of this disorder prevents the determination of the structure. The detailed nature of the disorder of the RC crystals has not been determined.

The order of the RC crystals was improved by exchanging the detergent. Crystals having form B6 were placed in a dialysis cell (see Methods) and the detergent LDAO was exchanged

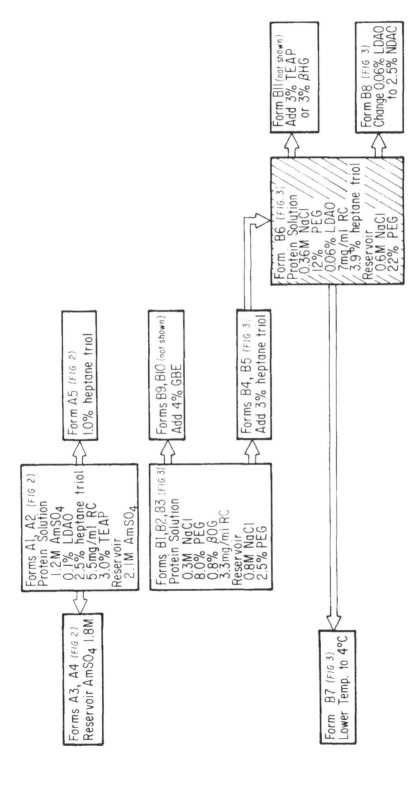

FIGURE 7.   Chart showing the effect of varying crystallization conditions on the observed crystal form. Optimal growth conditions resulting in crystal form B6 are shown stippled. This form was used to obtain the three dimensional structure of RCs from *Rhodobacter sphaeroides* R-26 to a resolution of 2.8 Å.[22-26] Arrows to the right indicate a change in detergent or amphiphile; arrows to the left a change in salt concentration or temperature.

TABLE 1
Different Crystal Forms of RCs From *Rhodobacter sphaeroides*, Their Space Groups, and Unit Cell Parameters[a]

Crystal form	Space group	a(Å)	b(Å)	c(Å)	β	RCs (number per unit cell)	RCs (number per asymmetric unit)	V_m (Å³/Da)
A1	C222	185	190	105	/	8	1	2.7
B2	P2	70	105	85	106°	2	1	2.1
B4	P2	142	138	143	106°	8	4	2.3
B5	$P2_12_12_1$	143	326	143	/	16	4	2.7
B6	$P2_12_12_1$	138.0	77.5	141.8	/	4	1	2.5

[a] The parameters for form B6 are for crystals after detergent exchange by dialysis (see text). The crystal volume per unit molecular mass, V_m, was calculated using 150,000 Da for the molecular mass of the RC and bound detergent.

Modified from Table 1 of Reference 12.

with 1.6% βOG (the concentrations of the other components of the mother liquor were matched in the reservoir solution). After 2 to 4 d of dialysis no disorder of the diffraction pattern was seen. This improvement was achieved only under specific conditions. Use of different βOG concentrations or other detergents either did not improve or worsened the ordering. In addition to the elimination of the additional reflections, the unit cell parameters of the dialyzed crystals were altered. The unit cell dimensions changed from 142.4, 75.5, and 141.8 Å to 138.0, 77.5, and 141.8 Å for a, b, and c, respectively.

The arrangement of all RCs in the crystal was determined from an analysis of the X-ray diffraction data. For form B6, the space group is $P2_12_12_1$, with four RCs per unit cell. The contacts between RCs involve only the hydrophilic regions. Detergent molecules of LDAO bound to the RC near the hydrophilic region may perturb the contacts between RCs. Exchange of LDAO with βOG could improve these contacts either due to the shorter chain length or because βOG does not bind to the RC near the contact region. This may explain the observed improvement in crystalline order after the detergent was exchanged. Detergent molecules that are bound away from the contact region should not contribute to the disorder as the distance between RCs is too far to allow contact between these detergent molecules. Except for two ordered detergent molecules bound to the RC,[28] the detergent molecules in the crystal appear to be too disordered to be observed in the electron density. The lack of structural information concerning the detergent prevents a quantitative analysis of the cause of disorder.

E. ROLE OF DETERGENTS AND AMPHIPHILES

During purification of membrane proteins, the lipids surrounding the protein are usually replaced with detergent; this often destabilizes the protein. For crystallization, the protein must be stable for several weeks; thus the choice of detergent is critical. In addition, we have found that amphiphiles can strongly influence the solubility of the protein. The RC, prepared with the detergent LDAO, was found to be very insoluble when ammonium sulfate was used as the salting agent. However, with the addition of the amphiphile heptane triol the RC was fully soluble at concentrations needed for crystallization. This demonstrates that the choice of amphiphile should be carefully considered when attempting to crystallize a membrane protein.

Due to the presence of detergent and amphiphiles, the crystallization process of membrane proteins possesses several new, independent parameters. As described in the previous

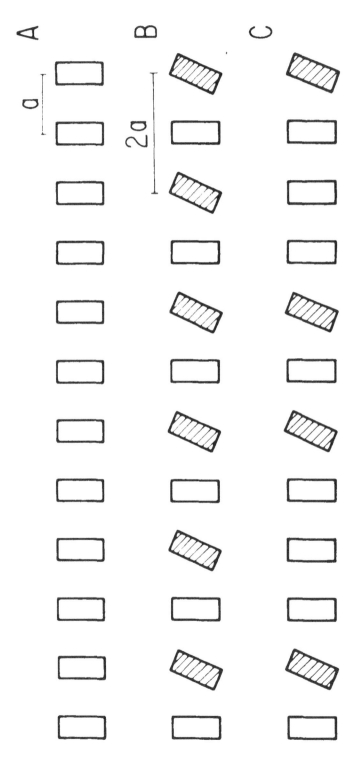

FIGURE 8. Schematic illustration of short range orientational disorder. Case A: crystal with unit cell length a. Case B: crystal with unit cell length 2a. Case C: disordered crystal with an effective unit cell length 2a.

sections, the influence of these new, nontraditional parameters is diverse and critical to the process of crystallization. Since the process of crystallization is complex, it is more easily understood by dividing it into three temporal phases[33] and examining the influence of these parameters in each temporal phase.

1. Nucleation

The critical stage of growing crystals of a new protein involves the attainment of conditions that lead to the successful nucleation of proteins. This stage is energetically unfavorable (i.e., a potential barrier must be overcome) and often leads to amorphous precipitation instead of crystallization. The aggregation of proteins is achieved by slowly increasing the protein concentration beyond the solubility limit. In the crystallization of the RC, this was achieved by removing water from the protein solution by vapor diffusion. Under several conditions discussed in this paper the detergent (and RC) phase separates from the PEG. Crystals often grew at the phase boundary (see Figure 4). This phase separation may provide a mechanism for the nucleation of the crystals.

2. Post Nucleation Growth

Once small crystals of a protein are observed, it is necessary to increase their size to obtain high resolution diffraction data. We have found that parameters involving the detergents and amphiphiles (e.g., type of detergent, length of carbon chain, concentrations of detergent and amphiphiles) influence the growth of crystals. In addition, alteration of one of these parameters was often useful in changing the space group of the crystals. This provided a procedure for obtaining large crystals and a desirable space group for the determination of the structure by X-ray diffraction.

3. Cessation of Growth

After a time interval of several weeks, the growth of the RC crystals ceased even if the protein concentration in the supernatant was increased. This cessation of growth may be due to the accumulation of errors (local disorder) occurring in the crystallization process.[33] One form of the RC crystals, form B6, was found to possess a short range disorder that may have contributed to the cessation of growth. Exchange of the detergent provided an improved packing arrangement that eliminated this disorder. This demonstrates the importance of the detergent and amphiphile chain length in establishing well-ordered crystals. It supports the hypothesis that the addition of small detergents and amphiphiles allows a better packing arrangement of the protein crystals.[38] It also suggests that the improvement of crystal size and order of other membrane proteins may be achieved by changing the length and type of detergent.

ACKNOWLEDGMENTS

We thank E. Abresch for the preparation of the RCs and D. C. Rees, H. Komiya, A. Chirino, and T. O. Yeates for helpful discussions. This work was supported by grants from the National Institutes of Health (AM 36053 and GM 13191).

REFERENCES

1. **Emerson, R. and Arnold, W.,** *J. Gen. Physiol.,* 16, 191, 1932.
2. **Duysens, L. N. M.,** Ph.D. Thesis, University of Utrecht, 1952.
3. **Govindjee, Ed.,** *Photosynthesis: Energy Conversion by Plants and Bacteria,* Academic Press, NY, 1982.
4. **Michel-Beyerle, M. E., Ed.,** *Antennas and Reaction Centers of Photosynthetic Bacteria,* Springer-Verlag, Berlin, 1985.
5. **Breton, J. and Vermeglio, A., Eds.,** *The Photosynthetic Bacterial Reaction Center,* Plenum Press, NY, 1988.
6. **Feher, G. and Okamura, M. Y.,** *The Photosynthetic Bacteria,* Clayton, R. K. and Sistrom, W. R., Eds., Plenum Press, NY, 1978, 349.
7. **Michel, H. and Oesterhelt, D.,** *Proc. Natl. Acad. Sci. U.S.A.,* 77, 1283, 1980.
8. **Garavito, R. M. and Rosenbusch, J. P.,** *J. Cell. Biol.,* 86, 327, 1980.
9. **Michel, H.,** *J. Mol. Biol.,* 158, 567, 1982.
10. **Garavito, R. M., Jenkins, J. A., Neuhaus, J. M., Pugsley, A. P., and Rosenbusch, J. P.,** *Ann. Microbiol.,* 133A, 37, 1982.
11. **Allen, J. P. and Feher, G.,** *Proc. Natl. Acad. Sci. U.S.A.,* 81, 4795, 1984.
12. **Feher, G. and Allen, J. P.,** *Molecular Biology of the Photosynthetic Apparatus,* Cold Spring Harbor Laboratories, NY, 1985, 163.
13. **Chang, C. H., Schiffer, M., Tiede, D., Smith, U., and Norris, J.,** *J. Mol. Biol.,* 186, 201, 1985.
14. **Allen, J. P., Feher, G., Yeates, T. O., Rees, D. C., Deisenhofer, J., Michel, H., and Huber, R.,** *Proc. Natl. Acad. Sci. U.S.A.,* 83, 8589, 1986.
15. **Ducruix, A. and Reiss Husson, F.,** *J. Mol. Biol.,* 193, 419, 1987.
16. **Frank, H. A., Taremi, S. S., and Knox, J. R.,** *J. Mol. Biol.,* 198, 139, 1987.
17. **Deisenhofer, J., Epp, O., Miki, K., Huber, R., and Michel, H.,** *J. Mol. Biol.,* 180, 385, 1984.
18. **Deisenhofer, J., Epp, O., Miki, K., Huber, R., and Michel, H.,** *Nature,* 318, 618, 1985.
19. **Michel, H., Epp, O., and Deisenhofer, J.,** *EMBO J.,* 5, 2445, 1986.
20. **Allen, J. P., Feher, G., Yeates, T. O., Rees, D. C., Eisenberg, D. S., Deisenhofer, J., Michel, H., and Huber, R.,** *Biophys. J.,* 49, 583a (abstr.), 1986.
21. **Chang, C. H., Tiede, D., Tang, J., Smith, U., Norris, J., and Schiffer, M.,** *FEBS Lett.,* 205, 82, 1986.
22. **Allen, J. P., Feher, G., Yeates, T. O., Rees, D. C., Deisenhofer, J., Michel, H., and Huber, R.,** *Proc. Natl. Acad. Sci. U.S.A.,* 83, 8589, 1986.
23. **Allen, J. P., Feher, G., Yeates, T. O., Komiya, H., and Rees, D. C.,** *Proc. Natl. Acad. Sci. U.S.A.,* 84, 5730, 1987.
24. **Allen, J. P., Feher, G., Yeates, T. O., Komiya, H., and Rees, D. C.,** *Proc. Natl. Acad. Sci. U.S.A.,* 84, 6162, 1987.
25. **Yeates, T. O., Komiya, H., Rees, D. C., Allen, J. P., and Feher, G.,** *Proc. Natl. Acad. Sci. U.S.A.,* 84, 6438, 1987.
26. **Allen, J. P., Feher, G., Yeates, T. O., Komiya, H., and Rees, D. C.,** *The Photosynthetic Bacterial Reaction Center,* Breton, J. and Vermeglio, A., Eds., Plenum Press, 1988, 5.
27. **Tiede, D. M., Budil, D. E., Tang, J., Kabbani, O. E., Norris, J. R., Chang, C. H., and Schiffer, M.,** *The Photosynthetic Bacterial Reaction Center,* Breton, J. and Vermeglio, A., Eds., Plenum Press, 1988, 13.
28. **Yeates, T. O., Komiya, H., Chirino, A., Rees, D. C., Allen, J. P., and Feher, G.,** *Proc. Natl. Acad. Sci. U.S.A.,* 85, 7993, 1988.
29. **Allen, J. P., Feher, G., Yeates, T. O., Komiya, H., and Rees, D. C.,** *Proc. Natl. Acad. Sci. U.S.A.,* 85, 8487, 1988.
30. **Komiya, H., Yeates, T. O., Rees, D. C., Allen, J. P., and Feher, G.,** *Proc. Natl. Acad. Sci. U.S.A.,* 85, 9012, 1988.
31. **Rees, D. C., Komiya, H., Yeates, T. O., Allen, J. P., and Feher, G.,** *Annu. Rev. Biochem.,* 58, 607, 1989.
32. **McPherson, A.,** *Preparation and Analysis of Protein Crystals,* John Wiley & Sons, NY, 1982.
33. **Feher, G. and Kam, Z.,** *Meth. Enzymol.,* 114, 77, 1985.
34. **Butler, W. F., Calvo, R., Fredkin, D. R., Isaacson, R. A., Okamura, M. Y., and Feher, G.,** *Biophys. J.,* 45, 947, 1984.
35. **Zeppezauer, M., Eklund, H., and Zeppezauer, E.,** *Arch. Biochem. Biophys.,* 126, 564, 1968.
36. **DeGiorgio, V. and Conti, M., Eds.,** *Physics of Amphiphiles, Micelles, Vesicles, and Microemulsions,* Elsevier/North-Holland Biochemical Press, Amsterdam, Netherlands, 1985.
37. **Zulauf, M.,** *Physics of Amphiphiles, Micelles, Vesicles and Microemulsions,* DeGiorgio, V. and Conti, M., Eds., Elsevier/North-Holland Biochemical Press, Amsterdam, Netherlands, 1985, 663.

38. **Michel, H.,** *Trends in Biological Science,* 1983, 56.
39. **Matthews, B. W.,** *J. Mol. Biol.,* 33, 491, 1968.
40. **Harburn, G., Taylor, C. A., and Welberry, T. R.,** *Atlas of Optical Transforms,* Bell, London, 1975.
41. **Welberry, T. R. and Galbraith, R.,** *J. Appl. Crystallogr.,* 6, 87, 1973.

Chapter 8

CRYSTALLIZATION OF THE LIGHT-HARVESTING CHLOROPHYLL A/B PROTEIN COMPLEX FROM CHLOROPLAST MEMBRANES

Werner Kühlbrandt

TABLE OF CONTENTS

I. INTRODUCTION

The light-harvesting chlorophyll a/b protein complex of photosystem II (LHC II) is the main antenna capturing solar energy in the chloroplasts of green plants and in green algae. It occurs predominantly in stacks of closely appressed photosynthetic membranes, the chloroplast grana. Excitation energy is passed to the reaction centers of photosystems (PS) I and II which carry out the light reactions of photosynthesis. LHC II is by far the most abundant protein in chloroplast membranes[1] and thus one of the most abundant membrane proteins. While several photosynthetic membrane proteins from purple bacteria (see Chapters 3, 5, 6, and 7) have been crystallized, LHC II is so far the only protein from the photosynthetic membranes of green plants to form three-dimensional crystals.

Apart from its function as a molecular antenna for intercepting light energy, LHC II is chiefly responsible for membrane appression and grana formation. Adhesion of photosynthetic membranes is mediated by mono- and divalent cations that screen the negative surface charge on the complex.[2] Divalent cations are roughly $50 \times$ more effective in this process than monovalent cations. To restore the appression of membranes de-stacked *in vitro*, 2 to 3 mM $MgCl_2$ or 100 to 150 mM NaCl is required.[3] The formation of crystals with isolated LHC II occurs at the same salt concentrations and it seems likely that membrane stacking *in vivo* and crystallization *in vitro* involve similar physical interactions.

A third function of LHC II in the regulation of photosynthesis is closely related to its role in membrane appression. A portion of the LHC II apoprotein within the membrane can become phosphorylated by a specific membrane-bound kinase that responds to the local redox potential, which in turn depends on the differential energy turnover of PS I and PS II.[4] Upon phosphorylation the net charge at neutral pH on the complex becomes more negative. As the electrostatic repulsion between phosphorylated complexes increases, LHC II is thought to redistribute into the nonappressed stromal regions of the photosynthetic membrane where some of its excitation energy is absorbed by the PS I reaction center.[2,5]

A. MOLECULAR COMPONENTS OF LHC II

SDS polyacrylamide gels of isolated, purified LHC II usually show several closely spaced bands in the 23 to 27 kD molecular weight range, depending on the plant species from which the complex has been isolated. In pea, the major LHC II polypeptide runs at an apparent molecular weight of 24 kD, preceded by two fainter, minor bands at 23 and 23.5 kDa[6] (Figure 1). The various bands probably correspond to the products of a small family of highly conserved nuclear genes which encode very similar polypeptides.[7,8] Two different genes coding for the major polypeptide of pea LHC II indicated 98% homology[9,10] and molecular weights of 25,000 and 25,013, respectively.[11] The sequences of major and minor polypeptides of pea LHC II seem to be virtually identical.[12]

LHC II binds most of the chlorophyll b (Chl b) in plants, containing an almost equal number of Chl a and Chl b molecules, whereas the majority of the other Chl protein complexes of green plants contain only Chl a. The visible spectrum of the complex is dominated by the Chl absorption bands. LHC II solubilized in nonionic detergents such as *n*-octyl-β-D-glucopyranoside (OG) or Triton X-100 shows Soret bands at 436 nm (Chl a) and 474 nm (Chl b) with a shoulder at 482 nm due to carotenoid.[13,16] The Chl a and Chl b Q_y absorption bands at 672 and 652 nm are resolved in purified LHC II. Upon precipitation, the absorption maximum of the Chl a Q_y band is red-shifted to 676 nm. The position of this band is a sensitive measure of the intactness of the complex. A blue shift to 669 nm or below indicates the onset of irreversible denaturation.

Amino acid analysis and standard Chl determination have indicated that each 25 kDa polypeptide binds a total of 15 Chl molecules,[14] in agreement with other findings of 14.6 Chl per 25 kDa protein[13] and more than 10 Chl per polypeptide.[11] Earlier reports of 6 to 7

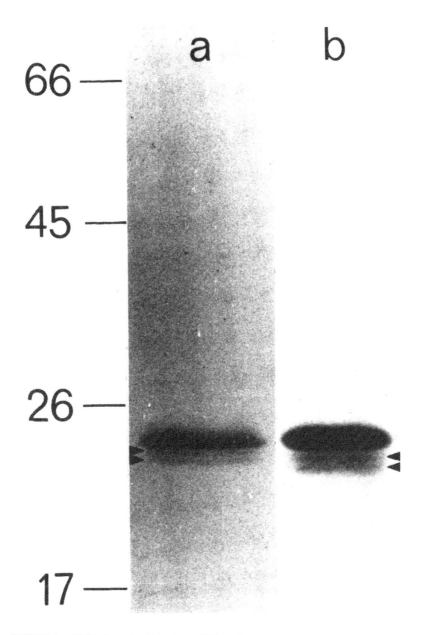

FIGURE 1. SDS/polyacrylamide gel of LHC II isolated from pea chloroplasts (a) and of a single octahedral crystal of pea LHC II. The position of molecular weight markers ($M_r \times 10^3$) is indicated for gel (a). Both tracks show a major polypeptide band at 24,000 apparent molecular weight and two minor bands at 23,000 and 23,500 apparent molecular weight. The heterogenous polypeptide composition of single crystals suggests that LHC II polypeptides are very similar and co-crystallize interchangeably.

Chl/polypeptide (see Reference 1) may have referred to LHC II that had lost some Chl due to treatment with harsh detergents, such as SDS. The ratio of Chl a/Chl b in octahedral crystals of LHC II, determined spectroscopically according to Arnon[15] is 1.15. This ratio is consistent with a total of 15 Chl and suggests that the monomeric complex binds 8 Chl a and 7 Chl b.

LHC II contains a variety of carotenoids (for a review see Reference 16). Ryrie et al.[17] found a molar ratio of total carotenoid to Chl of 0.282 in spinach LHC II. For each 15 chlorophylls there were 2.3 molecules of lutein, 1.3 of neoxanthin, 0.6 of violaxanthin, and 0.11 of β-carotene. These figures suggest nonstoichiometric, and therefore nonspecific binding of carotenoids. Lipid analysis of crystallizable LHC II indicated a molar ratio of 0.7 lipids/Chl.[6] The distribution of fatty acids was the same as in pea chloroplast lipids.

B. THE STRUCTURE OF LHC II

The structure analysis of two- and thin three-dimensional crystals of LHC II has shown that the complex is a trimer, consisting of three protein monomers related by threefold symmetry.[18,19] Negatively stained two-dimensional crystals of pea LHC II analyzed by electron microscopy and image processing yielded a three-dimensional map of the complex at 16 Å resolution.[18] The map showed two trimeric molecules in the hexagonal unit cell (a = 127 Å). The two complexes were arranged head-to-tail relative to one another. Each trimer was asymmetric with respect to the plane of the membrane, exposing a large area on one surface. The third dimension of the complex, judged from this map, was 60Å.

Electron diffraction of thin three-dimensional hexagonal plates of LHC II (see Section III.B.2) indicated the same unit cell in projection and the same symmetry of the diffraction pattern as found for two-dimensional crystals, suggesting that they were stacks of large, highly ordered two-dimensional crystals.[6] Image analysis of low-dose electron micrographs of unstained hexagonal plates revealed that the stacked two-dimensional crystals were not exactly in register but translationally displaced relative to one another. By computation it was possible to correct the images for the interference caused by these shifts. A 7 Å projection map was obtained.[19] Threefold symmetry of LHC II was consistent with the linear dichroism of hexagonal plates measured by microspectrophotometry.[20]

Sedimentation equilibrium ultracentrifugation of LHC II solubilized with 1% (wt/vol) OG have shown that the complex is a trimer not only in the crystalline state but also in detergent solution.[14] At very low protein concentration the complex dissociated into monomers and dimers. This demonstrated that the crystallizing molecule was the trimer and not the monomer. Picosecond time-resolved fluorescence and circular dichroism studies with LHC II in OG solution have indicated that the undisrupted trimer is essential for efficient energy transfer,[21] suggesting that LHC II is a trimer in its functional state *in vivo*.

II. ISOLATION AND FRACTIONATION OF LHC II

Because of its abundance in photosynthetic membranes of plants and green algae, LHC II is easy to isolate in quantities sufficient for crystallization experiments. The isolation procedure, though well documented[13,22] is described in some detail since experience has shown that isolation conditions are critical.

A. ISOLATION OF LHC II FROM CHLOROPLAST MEMBRANES

Isolation conditions used here refer to the preparation of LHC II from 2- to 4-week-old pea seedlings. The isolation of crystallizable LHC II from other plant tissues may require the adjustment of some parameters, in particular of the detergent/Chl used for solubilization. All steps were carried out at 4°C or on ice unless specified.

Chloroplasts corresponding to 30 mg Chl were isolated[23,24] and suspended in 10 mM NaCl, 5 mM EDTA, 1 mM Tricine pH 7.8. Osmotically disrupted chloroplasts were collected by centrifugation (10,000 × g, 10 min) and washed in the same buffer. Chloroplast membranes were destacked by suspending the pellet in 100 to 200 ml of 100 mM sorbitol, 5 mM EDTA pH 7.8. The pH was lowered to 6.0, adding drops of dilute HCl. The solution was stirred for 15 min at room temperature and centrifuged as above. The pellet was suspended in a known volume of 100 mM sorbitol. The total Chl in suspension was determined according to Arnon.[15] Membranes were pelleted by centrifugation as above and solubilized with 0.5%

(wt/vol) Triton X-100 (Boehringer, Mannheim, FRG) at a Chl concentration of 0.5 mg/ml. The suspension was stirred at 20°C for 30 min and clarified by centrifugation (40,000 × g, 30 min).

Eight milliliters of the dark green supernatant was layered on to 22 ml of a linear 0.1 to 1 M sucrose gradient in 0.05 to 0.1% (wt/vol) Triton X-100 and centrifuged (100,000 × g, 16 h) in a swing-out bucket or fixed-angle rotor. The gradients were examined against a small, directional incandescent light. The dark green band in the upper half of the gradient showed an intense red fluorescence characteristic of solubilized LHC II. The fluorescent band was collected (Chl a/Chl b ratio 1.5 to 1.85). Solid KCl was added to a final concentration of 300 mM and the solution was stirred for 10 to 15 min at room temperature. Precipitated LHC II was collected by centrifugation (25,000 × g, 10 min), washed twice in 100 mM KCl and once in distilled water by homogenization and centrifugation as above.

In order to grow three-dimensional crystals of LHC II, the complex was dissolved in n-nonyl-β-D-glucopyranoside (NG) at a final Chl concentration of 3 to 4 mg/ml by adding the appropriate amount of a 10% (wt/vol) NG and deionized water. The final concentration of NG was 0.7 to 0.9% (wt/vol). This stock solution was used fresh or stored at −20°C.

B. FRACTIONATION OF LHC II BY COLUMN CHROMATOGRAPHY

Isolated, detergent-solubilized LHC II was fractionated by ion-exchange column chromatography.[25] Having an isoelectric point between 4.0 and 4.55 (Reference 26) the complex is negatively charged at neutral pH and binds strongly to anion exchange resins. Pre-packed FPLC columns (Mono Q, Pharmacia, Milton Keynes, England) and columns packed with Fractogel TSK DEAE-650 (S) (BDH Chemicals, London, England) gave similar results. Columns were equilibrated with 10 mM Na/K phosphate or Tris-HCl buffer pH 7.0 containing 0.2% (wt/vol) Triton X-100 or 1.0% (wt/vol) OG. Better resolution of fractions was achieved with OG.

A 50 to 150 mM NaCl gradient in the same buffer released two main fractions at salt concentrations roughly equivalent to 100 and 120 mM NaCl. These were very similar in terms of absorption spectra and Chl a/Chl b ratio. Standard SDS polyacrylamide gel electrophoresis showed that the first major fraction was almost entirely free from minor LHC II polypeptides, whereas the second major fraction contained all three polypeptides normally found in isolated pea LHC II. The position of phosphorylated and nonphosphorylated LHC II in the elution profile was established by fractionating ^{32}P labeled LHC II. The radioactive label was enriched in several minor fractions which were shifted relative to the major fractions towards higher salt concentrations by roughly equal distances. With pea LHC II it was not possible to separate the phosphorylated complex because the major and minor fractions overlapped. However, with spinach LHC II the two major fractions eluted close together and the phosphorylated LHC II could be separated.[28] Surprisingly, none of the column fractions formed three-dimensional crystals. The reason for this is not clear. Treatment of the complex with high detergent/chlorophyll ratios, which was inevitable for column chromatography, may have weakened the interaction between the LHC II monomers, probably resulting in the breakdown of trimers into monomers and dimers which may be unstable under these conditions. It is also possible that column chromatography removes small molecules, such as lipids, from the complex which may be essential for the formation of three-dimensional crystals.

III. CRYSTALLIZATION OF LHC II

Isolated, detergent-solubilized LHC II crystallizes spontaneously in the presence of monovalent cations at concentrations above 100 mM or divalent cations at concentrations above 5 mM. Addition of salt to the solubilized complex results in the rapid formation of

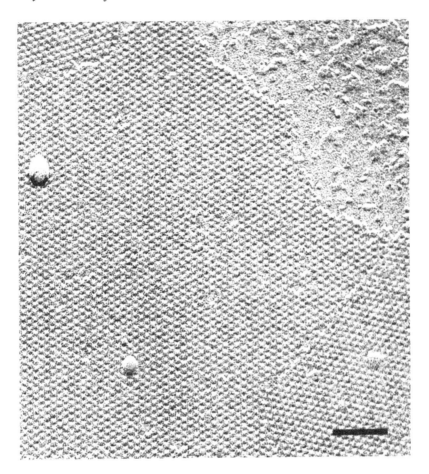

FIGURE 2.   Two-dimensional crystal of LHC II freeze-dried and shadowed unidirectionally with heavy metal. The shadow at the edge of the crystalline sheet indicates a thickness of 60 Å. (Scale bar, 100 nm.)

small, disordered microcrystalline aggregates. The problem in producing large, well ordered crystals is to slow down this spontaneous process.

A. TWO-DIMENSIONAL CRYSTALS

Although LHC II does not form regular arrays *in vivo*, the isolated complex can be induced, like other membrane proteins, to form "two-dimensional" membrane crystals in which the protein occupies positions on a two-dimensional lattice in the membrane plane. Such two-dimensional crystals are ideal objects for structure analysis by electron diffraction, electron microscopy and image processing.

Extensive, well-ordered two-dimensional crystals were obtained by dialysis of the isolated complex, solubilized in Triton X-100.[21] An aliquot of the fluorescent sucrose gradient fraction was precipitated by adding 4 M KCl to a final concentration of 300 mM. The precipitated complex was washed as described and finally dissolved in 0.2 volumes of 0.2% (wt/vol) Triton X-100 in deionized water. This solution was dialyzed against 300 mM KCl, 10 mM Tris HCl pH 7 in a water bath at 32 to 39°C. Dialysis was in glass tubes sealed at one end with dialysis tubing for convenient sampling during the experiment. Two-dimensional crystals measuring 2 to 10 μm (Figure 2) across formed within 30 to 60 min.

For structure analysis by electron microscopy, 5 μl drops were applied to carbon coated copper grids and stained with 2% (wt/vol) uranyl acetate containing 0.1% (wt/vol) Triton

X-100 for reproducible stain penetration. Other grids prepared at the same time were uni-directionally shadowed with heavy metal and examined in the electron microscope to check that the crystals were monolayers. Single layers greatly simplified the three-dimensional structure analysis by electron microscopy and image analysis.

B. THREE-DIMENSIONAL CRYSTALS

Three-dimensional crystals of LHC II can be grown from the complex solubilized in NG. Vapor diffusion at low salt concentrations yielded two forms, depending on conditions.[6]

1. Octahedra

To an aliquot of LHC II stock (3 to 4 mg/ml Ch, 0.7 to 0.9% (wt/vol) NG, 3 mM NaN$_3$), NG and crystallization buffer (75 mM KCl, 30 mM Na/K phosphate pH 6, 3 mM NaN$_3$) were added to a final concentration of 2.3 mg/ml Chl, 0.55 to 0.65% (wt/vol) NG, 25 mM KCl, 10 mM phosphate pH 6 and 3 mM NaN$_3$, corresponding to molar ratios of 6.9 to 8.3 NG/chlorophyll or 103 to 125 NG/monomer. The solution was mixed rapidly, left for 10 to 20 min at room temperature and centrifuged at low speed. Vapor diffusion was in sitting or hanging drops of darkly colored supernatant against 140 mM KCl and 3 mM NaN$_3$. Glass surfaces in contact with the LHC II solution were siliconized.

Small crystals appeared occasionally within minutes, but more usually overnight, and continued to grow for several days or weeks to a maximum size of 0.5 to 0.7 mm (body diagonal) at temperatures ranging from 4° to room temperature. The largest specimens formed at 16 to 18°C, at which temperature they seemed to be stable for several months. Octahedral crystals were obtained using phosphate buffers ranging from pH 4.5 to pH 9.5 without any apparent effect on crystal shape, although crystals grown at pH 6 to pH 7 tended to be larger. They were very dark green or black and showed a faint, isotropic red fluorescence when examined in incandescent white light. Crystals frequently appeared triangular or hex-agonal in outline (Figure 3), having shapes that derived from regular octahedra which indicated growth in the (1,1,1) direction. Fully developed octahedra were comparatively rare. Crystals had sharp edges and plane, well-developed faces. They were not birefringent.

Octahedral crystals mounted in mother liquor diffracted X-rays to 15 Å resolution. Attempts to harvest crystals into a protein-free buffer or into fresh mother liquor resulted in complete loss of internal order, as judged by X-ray diffraction, although to the eye they appeared undamaged. The X-ray diffraction pattern indicated a very large, cubic unit cell (a = 390 Å), containing up to several hundred LHC II monomers.[6] The octahedral habit and the lack of birefringence were consistent with a cubic unit cell.

2. Hexagonal Plates

A second type of three-dimensional LHC II crystals which grew as thin hexagonal plates formed less readily than octahedral crystals. Although both types grew under similar con-ditions, they rarely occurred in the same drop. The Chl concentration was 2.0 to 2.1 mg/ml at a detergent/protein ratio of 8.0 to 10.3 NG/Chl, or 120 to 153 NG/LHC II monomer. The reservoir was 80 to 120 mM Na/K phosphate, 3 mM phosphate, 3 mM NaN$_3$. Otherwise, crystallization conditions were the same as for octahedral crystals.

Bright green, perfectly hexagonal crystals appeared within several days, typically on the glass and on the surface, but occasionally throughout the drop. The crystals measured up to 200 μm across but never more than a few micrometers in thickness (Figure 4). They were easily distinguished from the octahedral form by their birefringence which was most pronounced edge-on or at glancing angles of 30 to 40° but absent face-on, and by a vivid, directional red fluorescence at the edges.

Hexagonal plates were too thin for X-ray crystallography, but electron diffraction of very thin specimens which formed within a day, with an estimated thickness of roughly 100

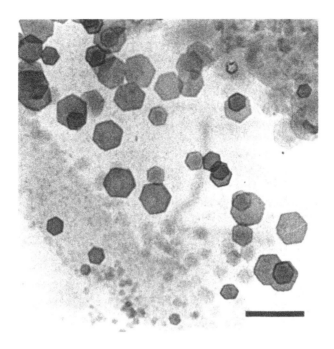

FIGURE 3. Octahedral crystals of LHC II. Crystals of this type measure up to 0.7 nm across and diffract X-rays to 15 Å resolution. (Scale bar, 0.2 mm.)

FIGURE 4. Hexagonal plates of LHC II. Crystals of this type grow to approximately 200 μm in diameter but measure only a few micrometers in thickness. Thin specimens diffract electrons to high resolution. (Scale bar, 0.2 mm.)

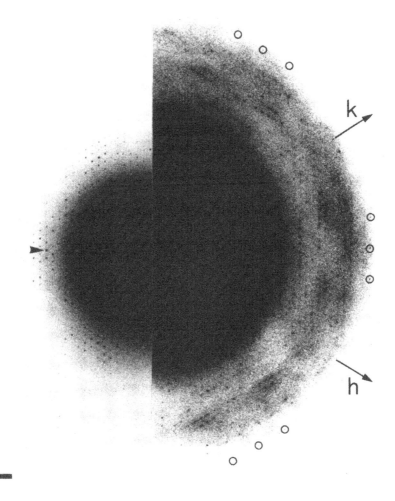

0.1 Å⁻¹

FIGURE 5. Electron diffraction pattern of a very thin hexagonal plate of LHC II. The symmetry of the diffraction pattern and the unit cell in projection are the same as for negatively stained crystalline sheets, suggesting that the hexagonal plates consist of stacked two-dimensional crystals. Sharp reflections at 3.7 Å resolution (circled) can be distinguished. Strong spots at 7 to 8 Å resolution around the (8,8) reflection (arrowhead) suggest α-helices perpendicular to the crystal plane.

nm, showed sharp reflection at 3.7 Å resolution (Figure 5). For structural studies by electron microscopy and electron diffraction, crystals were applied to a carbon-coated copper grid in a drop of mother liquor and washed with 0.5% (wt/vol) glucose or 0.5% (wt/vol) tannic acid which preserved their structure in the high vacuum of the electron microscope.

C. EFFECT OF DETERGENTS AND AMPHIPHILIC ADDITIVES

The choice of detergent was the most important single factor in obtaining three-dimensional crystals of LHC II. Judged by its absorption spectrum, the complex was stable for several weeks at 4°C in nonionic detergents such as OG, NG, Triton X-100, or n-dodecyl-β-D-maltoside (DM). Of these, NG yielded three-dimensional crystals reproducibly, OG occasionally, Triton X-100 and DM never. The highly specific requirement for NG may be an effect of micelle size and suggests that the NG-LHC II micelle fits best into the crystal lattice.

Zwitterionic detergents, such as lauryl-dimethyl-amineoxide (LDAO), its shorter chain analogues as well as *n*-octyl-2-hydroxy-ethylsulfoxide (OESO), *n*-octyl-rac-2,3-dihydroxypropyl sulfoxide (OPSO) and nonanoyl-*N*-methyl gucamide (MEGA-9) all denatured the complex. This was also true of small amphiphilic additives such as heptane-1,2,3-triol, benzamidine and methyl pentane diol. In the course of several days, these substances caused a gradual change of color from bright green to brownish green, indicating the breakdown of chlorophyll and therefore of the complex. After some time, small orange, highly birefringent needles or plates appeared. Electron diffraction indicated a unit cell of 24.8 Å × 11.1 Å × 6.2 Å, too small for a protein crystal.[6] The orange color of these needles suggested that they consisted of a carotenoid that had become detached from the complex.

LHC II was denatured in the same way by the two standard precipitants used in protein crystallization, ammonium sulfate and polyethylene glycol. When either substance was used instead of KCl/phosphate, irregular octahedral crystals formed within one day which soon dissolved as the complex changed color and orange microcrystals appeared. The reason for the surprising sensitivity of LHC II to these compounds is not known. Due to the large number of pigment molecules in the complex, LHC II probably has a rather open structure which may allow all but the least aggressive detergents and ions to invade and break down the native complex.

IV. CONCLUSIONS

LHC II, a trimeric integral membrane chlorophyll protein complex from the photosynthetic membrane of plant chloroplasts, can be crystallized from NG solution by vapor diffusion using methods similar to those developed for crystallizing bacteriorhodopsin,[27] the bacterial reaction centre complex from *Rhodopseudomonas viridis*[28] and matrix porin.[29] There are, however, several important points in which the crystallization of LHC II differs from that of other membrane proteins:

1. LHC II crystallizes spontaneously when salt is added to the detergent-solubilized complex.
2. LHC II forms crystals at considerably lower ionic strength than other membrane proteins crystallized to date.
3. LHC II is denatured by zwitterionic detergents and standard precipitants.
4. Small amphiphiles which promote the crystallization of other membrane proteins denature the complex.
5. Within the pH range of 4.5 to 9.5, the pH of the crystallizing solution is not critical to crystal formation.

Finally, it is interesting to note that the two crystal forms of LHC II described in this chapter seem to exemplify the two basic types of membrane protein crystals postulated by Michel.[30] Hexagonal plates, which are stacks of two-dimensional membrane crystals, seem to represent crystal type I with hydrophobic crystal contacts between proteins in the membrane plane and hydrophilic contacts between layers. It is difficult to increase both types of interaction simultaneously which may explain why hexagonal plates do not grow well in the third dimension. Since two-dimensional membrane crystals cannot be packed into a three-dimensional lattice having cubic symmetry, the octahedral crystal form seems to belong to Michel's crystal type II in which the forces between molecules are largely hydrophilic.

ACKNOWLEDGMENTS

I would like to thank Profs. J. Barber and D. M. Blow for their continued interest in this work. Financial support from the Agricultural and Food Research Council of Great Britain and a Heisenberg Research Fellowship (Deutsche Forschungsgemeinschaft) is gratefully acknowledged.

REFERENCES

1. **Thornber, J. P.**, Chlorophyll proteins: light-harvesting and reaction centre components of plants, *Annu. Rev. Plant Physiol.*, 26, 127, 1975.
2. **Barber, J.**, Influences of surface charges on thylakoid structure and function, *Annu. Rev. Plant Physiol.*, 33, 261, 1982.
3. **Izawa, S. and Good, N. E.**, Effect of salts and electron transport on the conformation of isolated chloroplasts. II. Electron microscopy, *Plant Physiol.*, 41, 544, 1966.
4. **Allen, J. F., Bennet, J., Steinback, K. E., and Arntzen, C. J.**, Chloroplast protein phosphorylation couples plastoquinone redox state to distribution of excitation energy between photosystems, *Nature*, 291, 25, 1981.
5. **Staehelin, L. A. and Arntzen, C. J.**, Regulation of chloroplast membrane function: protein phosphorylation changes the spatial organisation of membrane components, *J. Cell Biol.*, 97, 1327, 1983.
6. **Kühlbrandt, W.**, Three-dimensional crystals of the light-harvesting chlorophyll a/b-protein complex, *J. Mol. Biol.*, 194, 757, 1987.
7. **Karlin-Neumann, G. A., Komorn, B. D., Thornber, J. P., and Tobin, E. M.**, A chlorophyll a/b-protein encoded by a gene containing an intron with characteristics of a transposable element, *J. Mol. Appl. Genet.*, 3, 45, 1985.
8. **Dunsmuir, P., Smith, S. M., and Bedbrook, J.**, The major chlorophyll a/b-binding protein of petunia is composed of several polypeptides encoded by a number of distinct nuclear genes, *J. Mol. Appl. Genet.*, 2, 285, 1983.
9. **Coruzzi, G., Broglie, R., Cashmore, A., and Chua, N. H.**, Nucleotide sequences of two pea cDNA clones encoding the small subunit of ribulose 1,5-biphosphate carboxylase and the major chlorophyll a/b-binding thylakoid polypeptide, *J. Biol. Chem.*, 258, 1399, 1983.
10. **Cashmore, A. R.**, Structure and expression of a pea nuclear gene encoding a chlorophyll a/b-binding polypeptide, *Proc. Natl. Acad. Sci. U.S.A.*, 81, 2960, 1984.
11. **Li, J.**, Light-harvesting chlorophyll a/b-protein: three-dimensional structure of a reconstituted membrane lattice in negative stain, *Proc. Natl. Acad. Sci. U.S.A.*, 82, 386, 1985.
12. **Bürgi, R., Suter, F., and Zuber, H.**, Arrangement of the light-harvesting chlorophyll a/b-protein complex in the thylakoid membrane, *Biochim. Biophys. Acta*, 890, 346, 1987.
13. **Burke, J. J., Ditto, C. L., and Arntzen, C. J.**, Involvement of the light-harvesting complex in cation regulation of excitation energy distribution in chloroplasts, *Arch. Biochem. Biophys.*, 187, 252, 1978.
14. **Butler, P. J. G. and Kühlbrandt, W.**, Determination of the aggregate size in detergent solution of the light-harvesting chlorophyll a/b-protein complex from chloroplast membranes, *Proc. Natl. Acad. Sci. U.S.A.*, 85, 3797, 1988.
15. **Arnon, D. I.**, Copper enzymes in isolated chloroplasts. Polyphenol-oxidase in beta vulgaris, *Plant Physiol. (Bethesda)*, 24, 1, 1949.
16. **Siefermann-Harms, D.**, Carotenoids in photosynthesis. I. Location in photosynthetic membranes and light-harvesting function, *Biochim. Biophys. Acta*, 811, 325, 1985.
17. **Ryrie, I. J., Anderson, J. M., and Goodchild, D. J.**, The role of light-harvesting chlorophyll a/b-protein complex in chloroplast membrane stacking, *Eur. J. Biochem.*, 107, 345, 1980.
18. **Kühlbrandt, W.**, Three-dimensional structure of the light-harvesting chlorophyll a/b-protein complex, *Nature*, 307, 478, 1984.
19. **Kühlbrandt, W.**, The structure of light-harvesting chlorophyll-a/b-protein complex from chloroplast membranes at 7 Å resolution in projection, *J. Mol. Biol.*, 202, 849, 1988.
20. **Kühlbrandt, W., Becker, A., and Maentele, W.**, Chlorophyll dichroism of three-dimensional crystals of the light-harvesting chlorophyll a/b-protein complex *FEBS Lett.*, 226, 275, 1988.
21. **Ide, J., Klug, D., Kühlbrandt, W., Giorgi, L., and Porter, G.**, The state of detergent-solubilized light-harvesting chlorophyll a/b-protein complex as monitored by picosecond time-resolved fluorescence and circular dichroism, *Biochim. Biophys. Acta*, 893, 349, 1987.

22. **Kühlbrandt, W., Thaler, T., and Wehrli, E.**, The structure of membrane crystals of the light-harvesting chlorophyll a/b-protein complex, *J. Cell Biol.*, 96, 1414, 1983.

23. **Armond, P. A., Arntzen, C. J., Briantais, J.-M., and Vernotte, C.**, Differentiation of chloroplast lamellae: light-harvesting efficiency and grana development, *Arch. Biochem. Biophys.*, 175, 54, 1976.

24. **Nakatani, H. Y. and Barber, J.**, An improved method for isolating chloroplasts retaining their outer membranes, *Biochim. Biophys. Acta*, 461, 510, 1977.

25. **Kühlbrandt, W. and Barber, J.**, Separation of phosphorylated and non-phosphorylated light-harvesting chlorophyll a/b-protein complex by column chromatography, *Biochim. Biophys. Acta*, 934, 118, 1988.

26. **Mullet, J. E.**, The amino acid sequence of the polypeptide segment which regulates membrane adhesion (grana stacking) in chloroplasts, *J. Biol. Chem.*, 258, 9941, 1983.

27. **Michel, H.**, Three-dimensional crystals of a membrane protein complex. The photosynthetic reaction centre from *Rhodopseudomonas viridis, J. Mol. Biol.*, 158, 567, 1982.

28. **Michel, H. and Oesterhelt, D.**, Three-dimensional crystals of membrane proteins: bacteriorhodopsin, *Proc. Natl. Acad. Sci. U.S.A.*, 77, 1283, 1980.

29. **Garavito, R. M. and Rosenbusch, J. P.**, Three-dimensional crystals of an integral membrane protein: an initial X-ray analysis, *J. Cell Biol.*, 86, 327, 1980.

30. **Michel, H.**, Crystallization of membrane proteins, *Trends Biochem. Sci.*, 8, 56, 1983.

Chapter 9

THE CRYSTALLIZATION OF THE Ca^{2+}-ATPase OF SARCOPLASMIC RETICULUM

Anthony Martonosi, Kenneth A. Taylor, and Slawomir Pikula

TABLE OF CONTENTS

I. BACKGROUND

A. THE STRUCTURE OF Ca^{2+}-ATPase

The Ca^{2+}-transport ATPase of sarcoplasmic reticulum is a protein of 109,000 molecular weight.[1,2] About two thirds of its mass is exposed on the cytoplasmic surface of the bilayer,[3-6] where it interacts with ATP and Ca.[7-9] Hydrophobic transmembrane segments, representing about $^1/_3$ of the protein mass, anchor the Ca^{2+}-ATPase into the phospholipid bilayer and form the transmembrane channel for the translocation of Ca^{2+} (Reference 2).

The relative orientation of the cytoplasmic domains is presumably defined by intra- and intermolecular interactions, both in the regions exposed to the cytoplasm and in the hydrophobic interior of the bilayer. There are indications that the positions of the domains and the interactions between them are different in the two major conformations of the enzyme (E$_1$ and E$_2$) that alternate during Ca^{2+} transport.[10]

B. THE MECHANISM OF Ca^{2+}-TRANSPORT AND THE REQUIREMENT FOR PHOSPHOLIPIDS

The interaction of Ca^{2+} and ATP with the enzyme is followed by the formation of a phosphoenzyme intermediate (E$_1 \sim$ P) and the occlusion of Ca^{2+} in a form that is not accessible from either side of the membrane.[11] The E$_1 \sim$ P intermediate either donates its phosphate to ADP (ADP sensitive E $\sim$ P) forming ATP, or may be converted into E$_2 \sim$ P (ADP insensitive E $\sim$ P) with the eventual release of occluded Ca^{2+} into the lumen of the vesicle. The Mg^{2+}-dependent hydrolysis of the acylphosphate bond is followed by the isomerization of the enzyme from the E$_2$ into the E$_1$ form and a new cycle begins.[11]

The ATPase and Ca^{2+} transport activities of sarcoplasmic reticulum are dependent on phospholipids.[7] Reduction in the phospholipid content of the membrane by treatment with phospholipases A or C,[12-13] or by extraction with detergents,[14] causes inhibition of Ca^{2+} transport. After partial delipidation the ATPase activity and Ca^{2+} transport can be restored by readdition of phospholipids to the partially delipidated membrane.[12,15,16] A wide variety of neutral and anionic detergents can substitute for phospholipids in activating ATP hydrolysis, but cationic detergents are either inactive or inhibitory.[12,17] While they activate ATP hydrolysis, none of the currently available detergents can reactivate ATP-dependent Ca^{2+} transport in lipid depleted membranes. This implies either that vectorial translocation of Ca^{2+} by the ATPase specifically requires phospholipids, or, more likely, detergents cannot fulfill the barrier function of the phospholipid bilayer that is required for the retention and accumulation of transported Ca^{2+} on the trans side of the membrane.

There is general agreement that the inhibition of Ca^{2+} transport ATPase caused by partial delipidation from 100 to about 11 mol of phospholipid/mol of ATPase can be fully reversed by detergents or phospholipids.[16,17] The opinions are divided about the consequences of complete delipidation. According to Dean and Tanford[17] the Ca^{2+}-ATPase stripped of all but about 1 to 5 mol phospholipid/mol of ATPase can be fully reactivated by C$_{12}$E$_8$. According to Knowles et al.,[18] under similar conditions the reactivation did not exceed 50%; other investigators encountered total irreversible loss of ATPase activity after delipidation to phospholipid:ATPase mole ratios of less than 10.[14] Conclusions about the specific requirement for phospholipids in ATPase activity hinge upon these unexplained differences.

C. OLIGOMERS OF Ca^{2+}-ATPase—SOLUBILIZATION BY DETERGENTS

The concentration of Ca^{2+}-ATPase in sarcoplasmic reticulum of fast-twitch skeletal muscle is very high (20,000 to 30,000 ATPase molecules per μm^2 surface area) and there are good indications that in the native membrane a significant portion of the ATPase molecules are present in oligomeric form (for review see Reference 19). The Ca^{2+}-ATPase concentration in slow-twitch skeletal, cardiac and smooth muscles is significantly less than in fast-twitch muscles,[7] but even in these membranes the Ca^{2+}-ATPase is presumably in associated form.

The tendency for interaction between ATPase molecules in native and reconstituted membranes,[19-21] together with indirect evidence derived from kinetic studies,[22,23] led to speculations that the ATP-dependent Ca^{2+} transport involves dimers or perhaps higher oligomers of the Ca^{2+}-ATPase. Experimental test of this proposition was attempted by solubilization of the Ca^{2+}-ATPase in detergents under conditions where the dominant molecular species is expected to be the Ca^{2+}-ATPase monomer.[24] The interpretation of the data obtained on solubilized SR is complicated by the rapid irreversible loss of ATPase activity in detergent-solubilized preparations,[25] that is accompanied by structural changes in the Ca^{2+}-ATPase,[26] and the formation of enzymatically inactive ATPase aggregates that are not in equilibrium with the ATPase monomers.[27] Due to these complications, solubilized ATPase preparations usually contain a sizeable amount of denatured aggregated ATPase, and the equilibrium constant of ATPase-ATPase interactions determined by different investigators differs by 10^3.[27,28] Nevertheless, a consensus developed that detergent-solubilized Ca^{2+}-ATPase monomers catalyze reversibly all reaction steps, with the possible exception of the step involving the vectorial translocation of Ca^{2+}.

The lability of detergent-solubilized Ca^{2+}-ATPase is not surprising, since the dimensions of the transmembrane polypeptide segments, that are required for stability, presumably evolved to match the thickness of the phospholipid bilayer, that is quite different from the dimensions of detergent micelles. This implies that in detergent-solubilized systems, hydrophobic polypeptide segments destined to be immersed into the bilayer may be exposed to water, causing a rearrangement of the structure. Little systematic effort has been expended so far on the characterization of these structural changes. The prevention of these structural changes by rational choice of suitable detergents and conditions that provide long-term stability is an absolute requirement for the production of three-dimensional crystals of Ca^{2+}-ATPase that are suitable for X-ray diffraction analysis.

II. THE FORMATION OF TWO-DIMENSIONAL Ca^{2+}-ATPase CRYSTALS IN SARCOPLASMIC RETICULUM MEMBRANE

The Ca^{2+}-ATPase represents $\simeq 80\%$ of the protein content of sarcoplasmic reticulum.[7] Therefore sarcoplasmic reticulum vesicles can be used as starting material for the crystallization of Ca^{2+}-ATPase without need for prior purification. The crystallization of Ca^{2+}-ATPase in the native sarcoplasmic reticulum can be induced simply by stabilizing the enzyme either in the E_1 or in the E_2 conformation, using appropriate substrates and substrate analogs.

Vanadate (V) ions or inorganic phosphate, in the absence of calcium, shift the conformational equilibrium of Ca^{2+}-ATPase toward the E_2 state and induce the rapid formation of p2 type Ca^{2+}-ATPase crystals with Ca^{2+}-ATPase dimers as structural units.[3-5,29-34] Ca^{2+} or lanthanide ions interact with the Ca^{2+}-ATPase in the E_1 state.[35] The stabilization of the E_1 state gives rise to the spontaneous development of p1 type Ca^{2+}-ATPase crystals with Ca^{2+}-ATPase monomers as structural units.[10] The E_1 and E_2 type Ca^{2+}-ATPase crystals can be reversibly interconverted by appropriate changes in the Ca^{2+} concentration of the incubation medium.[10]

A. CRYSTALLIZATION OF Ca^{2+}-ATPase IN THE E_2 STATE BY VANADATE AND EGTA

Sarcoplasmic reticulum vesicles isolated from skeletal muscle homogenates by differential centrifugation[36] were suspended at 2°C in a medium of 0.1 M KCl, 10 mM imidazole, pH 7.4, 5 mM $MgCl_2$, 0.5 mM EGTA, and 5 mM Na_3VO_4 at a final protein concentration of 1 to 2 mg/ml. Crystalline arrays of Ca^{2+}-ATPase molecules develop within a few days on the surface of 60 to 70% of the vesicles in SR preparations obtained from fast-twitch skeletal muscle[29-33] and in less than 10% of the vesicles in preparations of slow-twitch skeletal or cardiac muscle.[34] Vesicles with extensive crystalline arrays usually acquire an elongated tubular shape. The vanadate-induced crystals consist of chains of Ca^{2+}-ATPase dimers wound in a right-handed helix around the tubules. Occasionally the crystals unwind, revealing isolated dimer chains that adhere to the support film.[31] The formation of the vanadate-induced crystals is promoted by inside-positive membrane potential, suggesting an influence of transmembrane potential on the conformation of the Ca^{2+}-ATPase.[32,37] The crystals are disrupted by Ca^{2+} at a concentration sufficient to saturate the high affinity Ca^{2+} binding sites of the Ca^{2+}-ATPase.[31] ATP also interferes with the crystallization.[31] The effects of Ca^{2+} and ATP are presumably due to the stabilization of the E_1 conformation.

Decavanadate binds to the Ca^{2+}-ATPase with higher affinity and stoichiometry than monovanadate and it is more potent in inducing its crystallization.[38-41] The Ca^{2+}-ATPase crystals formed by mono- or decavanadate are morphologically similar at the current level of resolution, but the decavanadate-induced crystals are less susceptible to disruption by $\approx$ 10^{-5} M Ca^{2+}.[41]

The E_2 type Ca^{2+}-ATPase crystals were analyzed by electron microscopy in negatively stained, freeze-fractured, and unstained frozen-hydrated preparations.

1. Observation of E_2 Type Ca^{2+}-ATPase Crystals by Negative Staining Electron Microscopy

The preservation of the structure by negative stain involves replacement of the aqueous regions of the sample with a nonvolatile salt of high atomic number. The image is an indirect representation of the structure, in which the protein and lipid regions are identified as regions of stain exclusion. Because the negative stains are water-soluble, the deposition of the stain is confined to the aqueous regions of the specimen, revealing their structure and delineating the boundary between the hydrophilic and the hydrophobic membrane domains.

For negative staining the crystalline vesicles were deposited on carbon coated parlodion films or on films of pure carbon and stained with 1 to 2% uranyl acetate or uranyl formate (pH 4.3); K-phosphotungstate (1% at pH 7.0) gave less satisfactory images.

After negative staining the vanadate-induced E_2 crystals of Ca^{2+}-ATPase display doublet rows of densities running at an angle of $\approx 57°$ to the long axis of the tubules (Figure 1). The doublet rows arise from chains of Ca^{2+}-ATPase dimers. The superposition of the structural features from the front and rear walls of the tubules frequently yields a diamond-like pattern.

The structural analysis of the crystals began with optical diffraction to determine the unit cell dimensions,[3] and was continued with three-dimensional reconstruction of the structure to the limit of resolution of about 25Å.[4] The unit cell dimensions of the E_2 type Ca^{2+}-ATPase crystals[3] are a = 65.9 Å, b = 114.4 Å, and γ = 77.9°. The two-sided plane group of the crystals is p2, with Ca^{2+}-ATPase dimers as structural units. The molecules in projection have a pear-shaped profile with a length of 66 Å and a width of 46 Å. Three-dimensional reconstruction from multiple tilted views of flattened, negatively stained crystalline tubules[42,43] provided information on the three-dimensional shape of the cytoplasmic domains of the molecule.[4]

The Ca^{2+}-ATPase molecules that form a dimer are connected with a bridge centered about 40 Å above the presumed cytoplasmic surface of the bilayer (Figure 2). The gap under the bridge is probably accessible to solutes. The orientation of the bridge is parallel to the long axis of the tubules, presumably because under these conditions the curvature of the tubule surface places little or no strain upon the structure of the dimer. Some density features were also observed in negatively stained preparations within the bilayer region of the membrane, but since both proteins and lipids are stain-excluding, the interpretation of these features required additional information from the analysis of frozen-hydrated specimens.

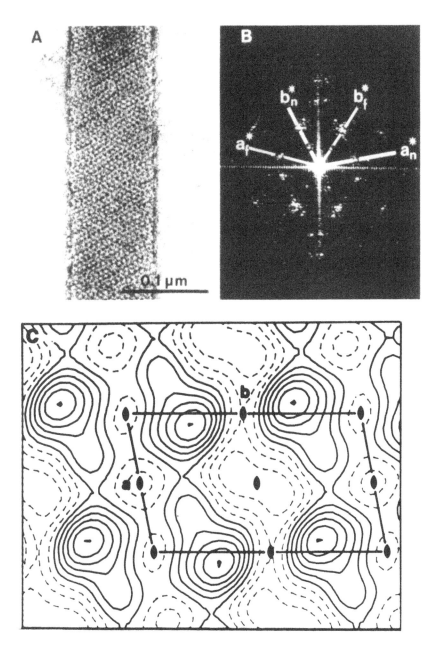

FIGURE 1. Image and optical diffraction pattern of vanadate-induced crystals. (A) Flattened tubules show superimposed images of the crystalline arrays on both sides. Doublet rows of stain-excluding densities are seen running at an angle of $57.6 \pm 2.1°$ to the tubule axis. Magnification $\times$ 222,000. The optical diffraction pattern (B) separates the contributions from the top and bottom sides; **a** and **b** represent the two reciprocal lattice vectors. (C) Projection map of mon-ovanadate-induced crystals. Map scale 0.55 mm/Å. (From Taylor, K. A., Dux, L., and Mar-tonosi, A., *J. Mol. Biol.*, 174, 193, 1984. With permission.)

2. Analysis of Unstained, Frozen-Hydrated Ca^{2+}-ATPase Crystals

Contrast in the case of unstained specimens is provided by the difference in electron density between proteins, lipids and the aqueous solution. The latter can be altered by the addition of various solutes. Electron microscopy of frozen specimens requires a stable cold specimen stage that can reach at least $-160°C$, and the use of low electron dose procedures.

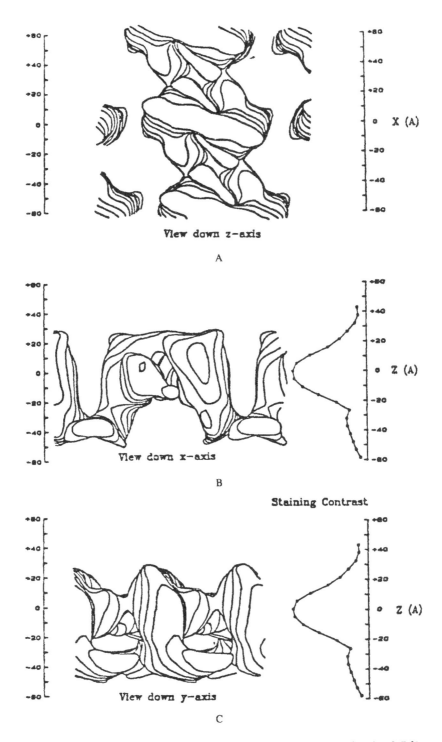

FIGURE 2.   Views of the three-dimensional structure of the cytoplasmic domain of Ca²⁺-ATPase in rabbit SR. (A) View down the z-axis. (B) The view down the x-axis indicates the height of the bridge above the surface of the bilayer and the space under the bridge that is penetrated by negative stain. The depth of the stain-filled grooves on the two sides of the dimer chain and the location of the lobes projecting from the main domain of the ATPase molecules toward the neighboring dimers are also resolved. (C) The view down the y-axis shows the inter-dimer connections through the lobes. (Reproduced from Taylor et al., Academic Press 1986a. With permission.)

The Ca^{2+}-ATPase crystals formed in a low salt medium (10 mM KCl) were spread on carbon film, glow discharged in amylamine, and frozen in liquid ethane slush, as described by Dubochet et al.[44] The grids were loaded under liquid nitrogen into a Philips EM 300 style cooling holder adapted for the Philips EM 400 electron microscope.[5] The optical diffraction data gave evidence for good preservation of both periodic structure and cylindrical shape.

Because the tubules are essentially helical cylinders, Fourier-Bessel reconstruction[45] was used to derive the three-dimensional structure of Ca^{2+}-ATPase from images of frozen-hydrated crystals.[5] The helical structure of the tubules is variable. One class that has been utilized for 3-D reconstruction has an axial repeat period of 358 Å and a helical structure of 80 ATPase dimers in 11 turns of a left-handed generic helix. The density features of unstained Ca^{2+}-ATPase tubule are contained in a cylindrical shell between 193 Å and 303 Å radius, and show the cytoplasmic and the intramembranous regions of the protein.[5] The cytoplasmic structures are similar to those obtained from negatively stained crystals. The most striking feature of the cytoplasmic surface is a deep helical groove between dimer chains, while on the luminal surface a deep groove extends down the center of the dimer chains. The intradimer bridge is centered at a 270 Å radius from the tubule axis, while the lobes that link the ATPase dimers into dimer chains are located at a radius of $\simeq$253 Å. Within the lipid bilayer region the main feature of the density map is a cylindrical domain about 40 Å in diameter, that extends through the bilayer at a slight angle. As a result, the intrabilayer domains of the Ca^{2+}-ATPase molecules within a dimer are farther apart on the luminal than on the cytoplasmic side of the bilayer. Only low density is present in the region running down the middle of the dimer chain. The dimer chains are linked together to form the surface lattice by a single connection at a radius of $\simeq$230 Å from the axis of the tubules. This connection is near the cytoplasmic surface of the lipid bilayer. The remarkable consistency of the structures derived from negatively stained and from unstained frozen-hydrated specimens gives confidence in the reality of the observed structures and demonstrates the complementarity of the two techniques.

3. Observation of E_2 Type Ca^{2+}-ATPase Crystals by Freeze-Etch Electron Microscopy

In vanadate-induced E_2 crystals of Ca^{2+}-ATPase, freeze-etch electron microscopy reveals regular arrays of oblique parallel ridges with spacings of $\simeq$110 Å on the P faces (cytoplasmic leaflet) and complementary grooves or furrows on the E faces (luminal leaflet).[46,47] In many instances the ridges break up into elongated particles of 95 Å length that repeat at 55 Å. When the direction of the shadow is almost parallel to the axis of the ridges the 95 Å particles can be resolved into two domains, which represent the intramembranous contacts between the two ATPase molecules of the two adjacent dimer chains. Complementary grooves on the E faces can also be resolved into rows of pits that match the particles of the ridges on the P faces. This structural description from freeze-etch electron microscopy supports the three-dimensional reconstruction derived from frozen-hydrated Ca^{2+}-ATPase crystals. Both methods demonstrate the existence of intramembranous contacts between dimer chains of ATPase molecules that give rise to the surface lattice. Such a structure is inconsistent with the existence of a transmembrane Ca^{2+} channel between the two ATPase molecules linked by the cytoplasmic bridge. The Ca^{2+} channel is probably a feature of each individual ATPase molecule, although the existence of a channel between the two ATPase molecules involved in the intramembranous contacts cannot be excluded. The significance of these contacts in Ca^{2+} translocation may be questionable, since all reaction steps of the Ca^{2+}-stimulated ATP hydrolysis with the exception of the actual translocation of Ca^{2+} can be demonstrated in detergent-solubilized sarcoplasmic reticulum containing the Ca^{2+}-ATPase in monomeric form.[24,25,27,48-52]

B. MEMBRANE CRYSTALS OF Ca^{2+}-ATPase INDUCED BY CALCIUM OR LANTHANIDES IN THE E_1 STATE

A different crystal form of the Ca^{2+}-ATPase develops in sarcoplasmic reticulum vesicles exposed at pH 8.0 to 10^{-5} to 10^{-4} M Ca^{2+} or 10^{-6} to 10^{-5} M lanthanide ions in a medium of 0.1 M KCl, 10 mM imidazole and 5 mM $MgCl_2$, for several days at 2°C.[10] Ca^{2+} and the lanthanides bind to the Ca^{2+}-ATPase in the E_1 form;[35] therefore these crystals are assumed to represent the structure and interactions of the Ca^{2+}-ATPase in the E_1 state. Analysis of the crystalline arrays by negative staining or freeze-fraction electron microscopy reveals obliquely oriented rows of particles corresponding to individual Ca^{2+}-ATPase molecules. The pear-shaped profiles of the cytoplasmic domains seen in projection are similar to those observed previously in vanadate-induced E_2 crystals. However, the two-sided plane group of the Ca^{2+}- or lanthanide-induced crystals is p1 and the unit cell dimensions are consistent with Ca^{2+}-ATPase monomers as structural units (Figure 3). The formation of the E_1 type crystals in the presence of Pr^{3+} is promoted by inside-negative potential. A similar crystal form is induced by 20 mM Ca^{2+} at pH 6.0.[53] While the E_1 crystals obtained at pH 8.0 are rather unstable[10] and disintegrate within a few days at 2°C, the crystals obtained at pH 6.0 with 20 mM Ca^{2+} can be preserved for several weeks without deterioration,[53] and may be more suitable for structural analysis of the Ca^{2+}-ATPase in the E_1 state.

Chelation of Ca^{2+} by EGTA causes the rapid disappearance of the E_1 crystals; subsequent addition of vanadate to this system induces the formation of E_2 type Ca^{2+}-ATPase crystals.[10] The reversible interconversion between the two crystal forms under conditions that are expected to stabilize either the E_1 or the E_2 conformation of the Ca^{2+}-ATPase further supports the notion that the two distinct crystal forms represent two distinct conformations of the enzyme.

By freeze-fracture electron microscopy the E_1 crystals display obliquely oriented rows of individual particles of about 60 Å diameter on the P face (cytoplasmic leaflet).[10,47] The ridges of particles seen on the P face of the E_2 crystals[46,47] are not evident in the E_1 crystals. In some areas particles are missing from the regular lattices and the spacings appear larger ($\approx$120 Å). Complementary oblique furrows on the convex E face (luminal leaflet), that are characteristic features of the E_2 crystals, are never observed on E_1 crystals, but obliquely oriented rows of pits are clearly visible. All these observations are consistent with the computer reconstructions of negatively stained E_1 type crystals induced by lanthanides.[10]

The three-dimensional reconstruction of the structure of Ca^{2+}-ATPase from E_1 type crystals is hindered by their infrequent occurrence and instability. Nevertheless, the structural information available so far clearly indicates substantial structural differences between the E_1 and E_2 conformations of Ca^{2+}-ATPase that affect the interactions between ATPase molecules, both in the cytoplasmic and in the intramembranous domains.

The following differences are most striking:

1. The cytoplasmic bridges that link together the ATPase dimers in the E_2 conformation are absent in the E_1 crystals.
2. The orientation of ATPase molecules with respect to the tubule axis is similar in the E_1 and in the E_2 crystals; this orientation persists in spite of the significant differences in the intermolecular contacts between nearest neighbor molecules in the two crystal forms.

It remains to be determined whether these structural differences between the E_1 and E_2 crystals are relevant to the structural transitions taking place during ATP-dependent Ca^{2+} transport. Based on circular dichroism measurements, the structural changes connected with the $E_1 \rightarrow E_2$ transition consist largely of hinge-like or relative sliding motion of domains,[26] but changes in the secondary structure of the Ca^{2+}-ATPase have also been inferred from Fourier transform infrared spectroscopy.[54]

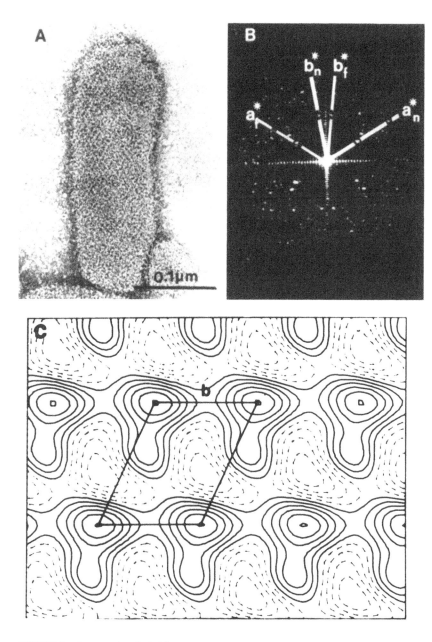

FIGURE 3. Image and optical diffraction pattern of praseodymium-induced crystals. (A) Crystallization was induced with 8 μM PrCl$_3$. Doublet tracks so prominent in vanadate-induced crystals are not evident in crystals induced with lanthanides. This results in an approximate halving of the b axis of the unit cell. Magnification × 222,000. (B) The image of the superimposed top and bottom lattices of the flattened cylinder give rise to two separate diffraction patterns. (C) Projection map of praseodymium-induced crystals. Map scale: 0.55 mm/Å. (From Dux, L., Taylor, K. A., Ting-Beall, H. P., and Martonosi, A., *J. Biol. Chem.*, 260, 11730, 1985. With permission.)

III. CRYSTALLIZATION OF Ca^{2+}-ATPase IN DETERGENT-SOLUBILIZED SARCOPLASMIC RETICULUM

Further advance toward a high resolution structure of Ca^{2+}-ATPase requires three-dimensional crystals of sufficient size and quality for X-ray diffraction analysis. A prerequisite for the formation of three-dimensional crystals is the solubilization of Ca^{2+}-ATPase from its membrane environment by detergents. Since the detergent-solubilized Ca^{2+}-ATPase is notoriously unstable, the first task was to find conditions that preserve the ATPase activity of solubilized enzyme for several months. It is likely that in all future studies involving relatively unstable membrane proteins, similar requirement for stabilization of the solubilized protein will be the crucial step toward crystallization.

By systematically testing several hundred conditions, we found that the Ca^{2+}-modulated ATPase activity is preserved for several months at 2°C under nitrogen in a crystallization medium of 0.1 M KCl, 10 mM K-MOPS, pH 6.0, 3 mM $MgCl_2$, 3 mM NaN_3, 5 mM dithiothreitol, 25 IU/ml Trasylol, 2 μg/ml 1,6-di-tert-butyl-p-cresol, 20 mM $CaCl_2$, 20% glycerol, 2 mg/ml sarcoplasmic reticulum protein, and 4 to 8 mg/ml of the appropriate detergent, such as $C_{12}E_8$, Brij 36T, Brij 56, or Brij 96. Under these conditions the sarcoplasmic reticulum membranes readily dissolve upon addition of the detergent and form a clear solution. Negative staining electron microscopy reveals individual Ca^{2+}-ATPase particles in various stages of aggregation and the absence of intact vesicles. After incubation for 6 to 10 d under nitrogen, ordered crystalline arrays begin to appear that increase in number and size during the next several weeks. The crystalline aggregates are fragile and can be easily fragmented by stirring or pipetting. The average diameter of the clusters is about 2000 to 10,000 Å, containing an estimated $\simeq 10^5$ to 10^6 molecules of the Ca^{2+}-ATPase.

Two distinct patterns are observed by electron microscopy of sectioned, negatively stained, and freeze-fractured specimens that are presumed to be different projections of the same structure.[54,55] In one view, layers of densities repeat at about 170 to 180 Å in negatively stained (Figure 4) and at 103 to 140 Å in sectioned (Figure 5) material; these represent side views of stacked lamellar arrays of ATPase molecules, probably interspersed with lipids and detergents. Between the layers there are 40 Å diameter particles that are reminiscent of the head portions of the Ca^{2+}-ATPase molecules. In the second view, the projected image of a single lamella or of several superimposed lamellae are seen (Figure 6) viewed down the z axis.

The purified Ca^{2+}-ATPase formed similar crystals to that obtained with sarcoplasmic reticulum, but at slightly lower detergent:protein weight ratios.

Analysis of electron micrographs obtained from edge-on views of these stacked sheets yield a sheet thickness of $\simeq 170$ Å and show lamellae with ATPase molecules extending out of both sides. In-plane projection of single sheets have been difficult to obtain, due to the strong tendency of the sheets to aggregate. Diffraction patterns, when obtained, have indicated that the unit cell is rectangular, with sides of 164.2 ± 2.2 and 55.5 ± 1.5 Å.

A. STABILIZATION OF THE SOLUBILIZED Ca^{2+}-ATPase

The conditions required for the crystallization of the Ca^{2+}-ATPase provided good stabilization of the solubilized enzyme with retention of ATPase activity over several months.[56] Of particular importance are the concentration of Ca^{2+}, the pH, and the presence of glycerol. Crystallization was impaired by changing the Ca^{2+} concentration at pH 6.0 from 20 to 5 mM, or by changing the pH at 20 mM Ca^{2+} concentration from 6.0 to either 5 or 8. Glycerol (20%) could be replaced with myoinositol, ethylene glycol, glucose, or sucrose, without significant changes in crystallization, but omission of polyhydroxy compounds increased the rate of inactivation of the ATPase and prevented the formation of three-dimensional crystals.

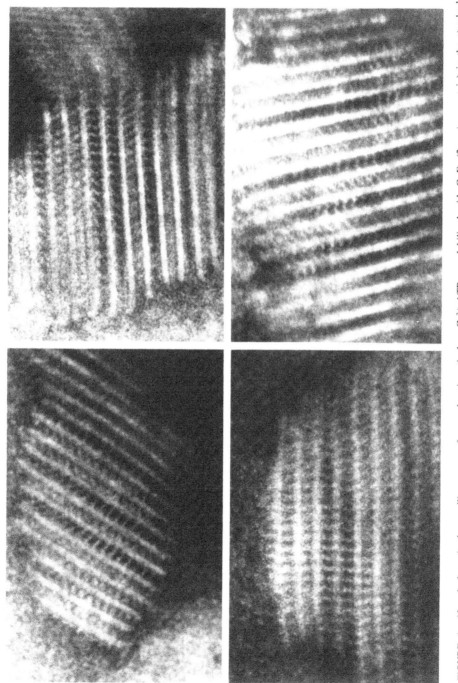

FIGURE 4. Negatively stained crystalline arrays of sarcoplasmic reticulum Ca^{2+}-ATPase solubilized with $C_{12}E_8$ (2 mg/mg protein) in the standard crystallization medium. Magnification × 308,000.

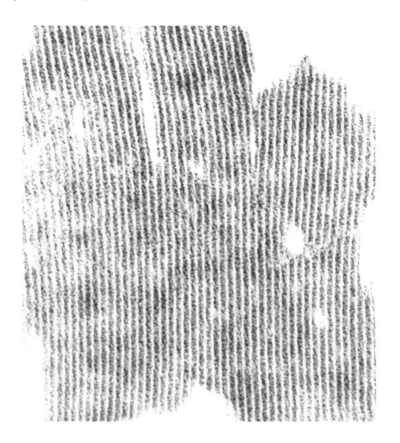

FIGURE 5.   Electron microscopy of Ca^{2+}-ATPase crystals in thin sections. Sarco-plasmic reticulum (2 mg protein/ml) was solubilized in the standard crystallization medium with $C_{12}E_8$ (2 mg/mg protein) and incubated under nitrogen at 2°C for 15 d. The crystalline sediment was embedded in Epon-Araldite mixture and processed for electron microscopy. Magnification × 207,000. Depending on conditions during fixation, embedding, sectioning and viewing the observed periodicities in different specimens varied between 103 and 147 Å.

B. SELECTION OF DETERGENTS

A total of 48 detergents were screened for crystallization and for preservation of Ca^{2+}-modulated ATPase activity, at detergent:protein ratios between 2:1 to 10:1. Optimal crystallization was observed with Brij 36T, Brij 56, and Brij 96 at a detergent-protein ratio of 4:1 and with $C_{12}E_8$ at a detergent:protein ratio of 2:1. The four selected detergents are structurally related polyoxyethylene glycol ethers of relatively short alkyl chain length. We were not successful in achieving crystallization with the following detergents: cholate, de-oxycholate, glycocholate, taurocholate, CHAPS, *n*-octylglucoside, octyl-β-D-thioglucoside, *n*-dodecylglucopyranoside, dodecyl-D-maltoside, *N*-lauroyl-*N*-methyltaurine sodium, *N*-lau-roylsarcosine sodium, *N,N*-dimethyldodecylamine-*N*-oxide, dimethyloctadecylamine-*N*-ox-ide, Tergitol 7, cetylbromide, Span 20, Span 40, Span 60, Span 80, Triton X-100, Triton N-101, Triton X-102, Triton X-165, Triton X-207, Triton X-305, Triton DF-12, Triton CF-10, Triton B-1956, Triton X-15, Triton X-45, the Zwittergent series (3-08, 3-10, 3-12, 3-14 and 3-16), the Tween series (20, 40, 60, 80 and 85), Brij 35, and Brij 58. The small amphiphiles, 1,2,6-trihydroxyhexane and 1,2,3-heptanetriol, were also ineffective. Since under different test conditions these detergents may be useful in inducing crystallization, their testing will continue.

The crystals induced by 20 m*M* Ca^{2+} in detergent-solubilized sarcoplasmic reticulum

FIGURE 6. Negatively stained Ca^{2+}-ATPase crystals in sarcoplasmic reticulum solubilized with $C_{12}E_8$ (2 mg/mg protein) in the standard crystallization medium. Magnification $\times$ 308,000.

and in purified Ca^{2+}-ATPase preparations represent ordered arrays of Ca^{2+}-ATPase molecules. Similar structures were not observed either in extracted sarcoplasmic reticulum lipids or in lipid/detergent mixtures. Therefore, it is unlikely that the arrays would consist of phospholipid-cholesterol-detergent micelles of the types described by Lucy and Glauert[57] and others.[58] Nevertheless, a contribution by phospholipids and detergents to the organization of the lamellar arrays cannot be excluded, since after delipidation of sarcoplasmic reticulum by extraction with deoxycholate[17] a minimum lipid content of $\simeq 10$ mol of phospholipids/mol of Ca^{2+}-ATPase was usually retained. We assume that the core of the lamellae may contain a lipid/detergent phase into which the hydrophobic tail portions of the Ca^{2+}-ATPase molecules are symmetrically inserted; therefore, the lamellae would be covered on both surfaces by the hydrophilic head groups of ATPase molecules. The three-dimensional character of these structures is suggested by electron microscopy of thin-sectioned and freeze-fractured specimens that regularly show stacked discs or lamellae in various orientations. In the association of the lamellae into three-dimensional structures, interactions between the hydrophilic head portions of the ATPase molecules are assumed to play a role. The high concentration of Ca^{2+} (20 mM), together with the relatively low pH (pH 6.0) required for crystallization, may promote the head group interactions by minimizing charge repulsions. Therefore, the crystalline stacks are assumed to be stabilized by hydrophobic and polar interactions involving both proteins and phospholipids. These structures may be different from genuine crystals of membrane proteins in which phospholipids do not play a structural role, the primary force for crystallization is the interaction between hydrophilic regions of the proteins, and the detergents serve merely to coat the hydrophobic protein surfaces.[59]

The microcrystals described here are useful for electron microscope studies but they are unsuitable for X-ray diffraction analysis, because of their small size and fragility. Experiments are in progress to generate larger and more stable three-dimensional crystal forms from fully delipidated Ca^{2+}-ATPase using detergent-small amphiphile combinations.[59,60]

ACKNOWLEDGMENTS

The contribution of Ms. Lois Epstein, Drs. L. Dux, H. Ping Ting-Beall, Camillo Peracchia, Sandor Varga, Nandor Mullner, Tamas Keresztes, Miklos Vegh, Peter Csermely, Sandor Papp, and Istvan Jona to various phases of these studies is gratefully acknowledged.

Work on this chapter was supported by research grants from the National Institutes of Health, United States Public Health Service (AM 26545, GM 30598), the National Science Foundation of the USA (PCM 84-03679, INT. 86-17848), and the Muscular Dystrophy Association. Kenneth A. Taylor is an Established Investigator of the American Heart Association.

REFERENCES

1. **MacLennan, D. H., Brandl, C. J., Korczak, B., and Green, N. M.,** Amino-acid sequence of a Ca^{2+} + Mg^{2+}-dependent ATPase from rabbit muscle sarcoplasmic reticulum, deduced from its complementary DNA sequence, *Nature,* 316, 696, 1985.
2. **Brandl, C. J., Green, N. M., Korczak, B., and MacLennan, D. H.,** Two Ca^{2+}-ATPase genes: homologies and mechanistic implications of deduced amino acid sequences, *Cell,* 44, 597, 1986.
3. **Taylor, K. A., Dux, L., and Martonosi, A.,** Structure of the vanadate-induced crystals of sarcoplasmic reticulum Ca^{2+}-ATPase, *J. Mol. Biol.,* 174, 193, 1984.
4. **Taylor, K. A., Dux, L., and Martonosi, A.,** Three-dimensional reconstruction of negatively stained crystals of the Ca^{2+}-ATPase from muscle sarcoplasmic reticulum, *J. Mol. Biol.,* 187, 417, 1986.
5. **Taylor, K. A., Ho, M. H., and Martonosi, A.,** Image analysis of the Ca^{2+}-ATPase from sarcoplasmic reticulum, *Ann. N.Y. Acad. Sci.,* 483, 31, 1986.
6. **Blasie, J. K., Herbette, L. G., Pascolini, D., Skita, V., Pierce, D. H., and Scarpa, A.,** Time-resolved X-ray diffraction studies of the sarcoplasmic reticulum membrane during active transport, *Biophys. J.,* 48, 9, 1985.
7. **Martonosi, A. N. and Beeler, T. J.,** The mechanism of Ca^{2+} transport by sarcoplasmic reticulum, in *Handbook of Physiology. Skeletal Muscle,* Peachey, L. D. and Adrian, R. H., Eds., American Physiological Society, Bethesda, 1983, 417.
8. **Inesi, G. and de Meis, L.,** Kinetic regulation of catalytic and transport activities in sarcoplasmic reticulum ATPase, in *The Enzymes of Biological Membranes,* Vol. 3, Martonosi, A., Ed., Plenum, New York, 1985, 157.
9. **Inesi, G.,** Mechanism of calcium transport, *Annu. Rev. Physiol.,* 47, 573, 1985.
10. **Dux, L., Taylor, K. A., Ting-Beall, H. P., and Martonosi, A.,** Crystallization of the Ca^{2+}-ATPase of sarcoplasmic reticulum by calcium and lanthanide ions, *J. Biol. Chem.,* 260, 11730, 1985.
11. **de Meis, L. and Vianna, A. L.,** Energy interconversion by the Ca^{2+} dependent ATPase of the sarcoplasmic reticulum, *Annu. Rev. Biochem.,* 48, 275, 1979.
12. **Martonosi, A., Donley, J. R., and Halpin, R. A.,** Sarcoplasmic reticulum. III. The role of phospholipids in the adenosine triphosphatase activity and Ca^{2+} transport, *J. Biol. Chem.,* 243, 61, 1968.
13. **Martonosi, A., Donley, J. R., Pucell, A. G., and Halpin, R. A.,** Sarcoplasmic reticulum. XI. The mode of involvement of phospholipids in the hydrolysis of ATP by sarcoplasmic reticulum membranes, *Arch. Biochem. Biophys.,* 144, 529, 1971.
14. **Hidalgo, C., de la Fuente, M., and Gonzalez, M. E.,** Role of lipids in sarcoplasmic reticulum; a higher lipid content is required to sustain phosphoenzyme decomposition than phosphoenzyme formation, *Arch. Biochem. Biophys.,* 247, 365, 1986.
15. **Martonosi, A.,** Role of phospholipids in ATPase activity and Ca^{2+} transport of fragmented sarcoplasmic reticulum, *Fed. Proc.,* 23, 913, 1964.
16. **Bennett, J. P., McGill, K. A., and Warren, G. B.,** The role of lipids in the functioning of a membrane protein: the sarcoplasmic reticulum calcium pump, *Curr. Top. Membr. Transp.,* 14, 127, 1980.
17. **Dean, W. L. and Tanford, C.,** Properties of a delipidated detergent-activated Ca^{2+}-ATPase, *Biochemistry,* 17, 1683, 1978.

18. **Knowles, A. F., Eytan, E., and Racker, E.,** Phospholipid-protein interactions in the Ca^{2+}-adenosine triphosphatase of sarcoplasmic reticulum, *J. Biol. Chem.*, 251, 5161, 1976.

19. **Martonosi, A., Taylor, K. A., Varga, S., and Ting-Beall, H. P.,** The molecular structure of sarcoplasmic reticulum, in *Electron Microscopy of Proteins*, Vol. 6, Harris, J. R. and Horne, R. W., Eds., Academic Press, London, 1987, 255.

20. **Vanderkooi, J. M., Ierokomos, A., Nakamura, H., and Martonosi, A.,** Fluorescence energy transfer between Ca^{2+} transport ATPase molecules in artificial membranes, *Biochemistry*, 16, 1262, 1977.

21. **Papp, S., Pikula, S., and Martonosi, A.,** Fluorescence energy transfer as an indicator of Ca^{2+}-ATPase interactions in sarcoplasmic reticulum, *Biophys. J.*, 51, 205, 1987.

22. **Kurobe, Y., Nelson, R. W., and Ikemoto, N.,** Reversible control of oligomeric interaction of the sarcoplasmic reticulum calcium ATPase with the use of a cleavable cross-linking agent, *J. Biol. Chem.*, 258, 4381, 1983.

23. **Ikemoto, N. and Nelson, R. W.,** Oligomeric regulation of the later reaction steps of the sarcoplasmic reticulum calcium ATPase, *J. Biol. Chem.*, 259, 11790, 1984.

24. **Tanford, C.,** Twenty questions concerning the reaction cycle of the sarcoplasmic reticulum calcium pump, *CRC Crit. Rev. Biochem.*, 17, 123, 1984.

25. **Møller, J. V., Andersen, J. P., and le Maire, M.,** The sarcoplasmic reticulum Ca^{2+}-ATPase, *Mol. Cell. Biochem.*, 42, 83, 1982.

26. **Csermely, P., Katopis, C., Wallace, B. A., and Martonosi, A.,** The $E_1 \rightarrow E_2$ transition of Ca^{2+} transport occurs without major changes in secondary structure, *Biochem. J.*, 241, 663, 1987.

27. **Andersen, J. P., Vilsen, B., Nielsen, H., and Møller, J. V.,** Characterization of detergent-solubilized sarcoplasmic reticulum Ca^{2+}-ATPase by high-performance liquid chromatography, *Biochemistry*, 25, 6439, 1986.

28. **Silva, J. L. and Verjovski-Almeida, S.,** Monomer-dimer association constant of solubilized sarcoplasmic reticulum ATPase, *J. Biol. Chem.*, 260, 4764, 1985.

29. **Dux, L. and Martonosi, A.,** Two-dimensional arrays of proteins in sarcoplasmic reticulum and purified Ca^{2+}-ATPase vesicles treated with vanadate, *J. Biol. Chem.*, 258, 2599, 1983.

30. **Dux, L. and Martonosi, A.,** Ca^{2+}-ATPase membrane crystals in sarcoplasmic reticulum. The effect of trypsin digestion, *J. Biol. Chem.*, 258, 10111, 1983.

31. **Dux, L. and Martonosi, A.,** The regulation of ATPase-ATPase interactions in sarcoplasmic reticulum membranes. I. The effects of Ca^{2+}, ATP and inorganic phosphate, *J. Biol. Chem.*, 258, 11896, 1983.

32. **Dux, L. and Martonosi, A.,** The regulation of ATPase-ATPase interactions in sarcoplasmic reticulum membranes. II. The influence of membrane potential, *J. Biol. Chem.*, 258, 11903, 1983.

33. **Dux, L. and Martonosi, A.,** Membrane crystals of Ca^{2+}-ATPase in sarcoplasmic reticulum of normal and dystrophic muscle, *Muscle Nerve*, 6, 566, 1983.

34. **Dux, L. and Martonosi, A.,** Membrane crystals of Ca^{2+}-ATPase in sarcoplasmic reticulum of fast and slow skeletal and cardiac muscles, *Eur. J. Biochem.*, 141, 43, 1984.

35. **Jona, I. and Martonosi, A.,** The effects of membrane potential and lanthanides on the conformation of the Ca^{2+} transport ATPase in sarcoplasmic reticulum, *Biochem. J.*, 234, 363, 1986.

36. **Nakamura, H., Jilka, R. L., Boland, R., and Martonosi, A. N.,** Mechanism of ATP hydrolysis by sarcoplasmic reticulum and the role of phospholipids, *J. Biol. Chem.*, 251, 5414, 1976.

37. **Beeler, T. J., Dux, L., and Martonosi, A.,** The effect of Na_3VO_4 and membrane potential on the structure of sarcoplasmic reticulum, *J. Membr. Biol.*, 78, 73, 1984.

38. **Varga, S., Csermely, P., and Martonosi, A.,** The binding of vanadium (V) oligoanions to sarcoplasmic reticulum, *Eur. J. Biochem.*, 148, 119, 1985.

39. **Csermely, P., Varga, S., and Martonosi, A.,** Competition between decavanadate and fluorescein isothiocyanate on the Ca^{2+}-ATPase of sarcoplasmic reticulum, *Eur. J. Biochem.*, 150, 455, 1985.

40. **Csermely, P., Martonosi, A., Levy, G. C., and Ejchart, A. J.,** ^{51}V-n.m.r. analysis of the binding of vanadium (V) oligoanions to sarcoplasmic reticulum, *Biochem. J.*, 230, 807, 1985.

41. **Varga, S., Csermely, P., Mullner, N., Dux, L., and Martonosi, A.,** Effect of chemical modification on the crystallization of Ca^{2+}-ATPase in sarcoplasmic reticulum, *Biochim. Biophys. Acta*, 896, 187, 1987.

42. **Henderson, R. and Unwin, P. N. T.,** Three-dimensional model of purple membrane obtained by electron microscopy, *Nature (London)*, 257, 28, 1975.

43. **Amos, L. A., Henderson, R., and Unwin, P. N. T.,** Three-dimensional structure determination by electron microscopy of two-dimensional crystals, *Prog. Biophys. Mol. Biol.*, 39, 183, 1982.

44. **Dubochet, J., Lepault, J., Freeman, R., Berriman, J. A., and Homo, J. -C.,** Electron microscopy of frozen water and aqueous solutions, *J. Microsc. (Oxford)*, 128, 219, 1982.

45. **DeRosier, D. J. and Moore, P. B.,** Reconstruction of three-dimensional images from electron micrographs of structures with helical symmetry, *J. Mol. Biol.*, 52, 355, 1970.

46. **Peracchia, C., Dux, L., and Martonosi, A.,** Crystallization of intramembrane particles in rabbit sarcoplasmic reticulum vesicles by vanadate, *J. Muscle Res. Cell Motil.*, 5, 431, 1984.

47. **Ting-Beall, H. P., Burgess, F. M., Dux, L., and Martonosi, A.,** Microscopic analysis of two-dimensional crystals of the Ca^{2+}-transport ATPase—a freeze-fracture study, *J. Muscle Res. Cell Motil.,* 8, 252, 1987.

48. **Martin, D. W.,** Active unit of solubilized sarcoplasmic reticulum calcium adenosine triphosphatase: an active enzyme centrifugation analysis, *Biochemistry,* 22, 2276, 1983.

49. **Martin, D. W. and Tanford, C.,** Solubilized monomeric sarcoplasmic reticulum Ca pump protein. Phosphorylation by inorganic phosphate, *FEBS Lett.,* 177, 146, 1984.

50. **Martin, D. W., Tanford, C., and Reynolds, J. A.,** Monomeric solubilized sarcoplasmic reticulum Ca^{2+} pump protein: demonstration of Ca^{2+} binding and dissociation coupled to ATP hydrolysis, *Proc. Natl. Acad. Sci. U.S.A.,* 81, 6623, 1984.

51. **Andersen, J. P., Jorgensen, P. L., and Møller, J. V.,** Direct demonstration of structural changes in soluble monomeric Ca^{2+}-ATPase associated with Ca^{2+} release during the transport cycle, *Proc. Natl. Acad. Sci. U.S.A.,* 82, 4573, 1985.

52. **Vilsen, B. and Andersen, J. P.,** Occlusion of Ca^{2+} in soluble monomeric sarcoplasmic reticulum Ca^{2+}-ATPase, *Biochim. Biophys. Acta,* 855, 429, 1986.

53. **Dux, L., Pikula, S., Mullner, N., and Martonosi, A.,** Crystallization of Ca^{2+}-ATPase in detergent-solubilized sarcoplasmic reticulum, *J. Biol. Chem.,* 262, 6439, 1987.

54. **Arrondo, J. L. R., Mantsch, H. H., Mullner, N., Pikula, S., and Martonosi, A.,** Infrared spectroscopic characterization of the structural changes connected with the $E_1 \rightarrow E_2$ transition in the Ca^{2+}-ATPase of sarcoplasmic reticulum, *J. Biol. Chem.,* 262, 9037, 1987.

55. **Taylor, K. A., Mullner, N., Pikula, S., Dux, L., Peracchia, C., Varga, S., and Martonosi, A.,** Electron microscope observations on Ca^{2+}-ATPase microcrystals in detergent-solubilized sarcoplasmic reticulum, *J. Biol. Chem.,* 263, 5287, 1988.

56. **Pikula, S., Mullner, N., Dux, L., and Martonosi, A.,** Stabilization and crystallization of Ca^{2+}-ATPase in detergent-solubilized sarcoplasmic reticulum, *J. Biol. Chem.,* 263, 5277, 1988.

57. **Lucy, J. A. and Glauert, A. M.,** Structure and assembly of macromolecular lipid complexes composed of globular micelles, *J. Mol. Biol.,* 8, 727, 1964.

58. **De Kruijff, B., Cullis, P. R., Verkleij, A. J., Hope, M. J., Van Echteld, C. J. A., and Taraschi, F.,** Lipid polymorphism and membrane function, in *The Enzymes of Biological Membranes,* Vol. 1, 2nd ed., Martonosi, A., Ed., Plenum, New York, 1985, 131.

59. **Michel, H.,** Crystallization of membrane proteins, *Trends Biochem. Sci.,* 8, 56, 1983.

60. **Deisenhofer, J., Epp, O., Miki, K., Huber, R., and Michel, H.,** Structure of the protein subunits in the photosynthetic reaction centre of *Rhodopseudomonas viridis* at 3 Å resolution, *Nature,* 318, 618, 1985.

Chapter 10

TWO DIMENSIONAL CRYSTALS OF THE *RHODOPSEUDOMONAS VIRIDIS* PHOTOSYNTHETIC REACTION CENTER

Kenneth R. Miller

TABLE OF CONTENTS

I. INTRODUCTION

In recent years, the photosynthetic membrane of *Rhodopseudomonas viridis,* a purple nonsulfur photosynthetic bacterium, has emerged as a major system for the study of biological membranes and their polypeptides. The reasons for this have little to do with the photosynthetic membrane per se, and even less to do with the photosynthetic bacteria as a taxonomic group. Rather, the fact that these membranes and their components can be induced to form well-ordered two- and three-dimensional crystalline lattices has made them exceptionally useful as objects of crystallographic study. Earlier work from my laboratory reported two- and three-dimensional reconstructions[1,2] of the regular lattice of subunits which occur in the *Rhodopseudomonas viridis* photosynthetic membrane *in situ,* but this chapter will focus on artificially produced two-dimensional crystals produced from one of the membrane's major components: the photosynthetic reaction center.

II. THE *RHODOPSEUDOMONAS VIRIDIS* REACTION CENTER

Photosynthetic reaction centers are membrane-bound pigment-protein complexes which function in the primary reactions of photosynthesis. They contain a series of pigment molecules which are capable of absorbing energy from sunlight (as well as from accessory pigments outside the reaction center) and using that energy to drive a charge separation which supplies energized electrons to a membrane-bound electron transport chain. The *Rhodopseudomonas viridis* photosynthetic reaction center is a membrane protein consisting of four protein subunits. Three of these polypeptides are similar to those found in reaction centers from other photosynthetic bacteria, and are designated H (heavy), M (medium), and L (light) on the basis of their behavior in SDS-polyacrylamide gels. The *Rhodopseudomonas viridis* reaction center also contains a tightly bound cytochrome molecule, which supplies electrons to the primary electron donor after it has become oxidized.

A fundamental advance in the study of this protein took place in 1982 when Michel reported the formation of well-ordered three-dimensional crystals of purified reaction center which were suitable for analysis by X-ray diffraction.[3] The detailed structure of this polypeptide has now been reported by Michel and his associates.[4-6] This elegant model has made it possible for the first time to study the internal details of a membrane protein. The fact that the crystallized protein is a photosynthetic reaction center has been an added bonus for the study of energy-transducing membranes.

Although the X-ray map of the reaction center provides fundamental information regarding the internal details of this membrane protein, it is by no means assured that all membrane proteins will ultimately be capable of crystallization in a way that will yield the same degree of information. Therefore, it may be useful to summarize a series of experimental observations which have provided us with useful information about the *Rhodopseudomonas viridis* photosynthetic reaction center at the lower levels of resolution made possible by the use of the electron microscope to analyze sheetlike two-dimensional crystals of the same reaction center.

III. THE FORMATION OF TWO-DIMENSIONAL CRYSTALS

Shortly after Michel's report of the formation of crystals of the reaction center suitable for X-ray analysis,[3] we were able to induce purified reaction centers to form two-dimensional crystalline sheets with a high degree of order.[7] As reported in that communication, we used the detergent LDAO (*N,N*-dimethyldodecylamine-*N*-oxide) at a 1% concentration to solubilize isolated photosynthetic membranes. After binding the detergent-soluble protein fraction to a hydroxylapatite column, the reaction center protein was eluted by a linear salt gradient.

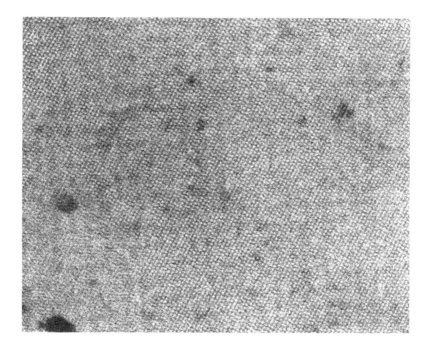

FIGURE 1. Two dimensional crystals of the *Rhodopseudomonas viridis* photosynthetic reaction center. Sheetlike crystals are formed by gradual dialysis of detergent-solubilized proteins at low pH as described.[7] The individual subunits visible in this negatively-stained preparation correspond to individual reaction centers. Magnification × 150,000.

Typically, the ratio OD_{280}/OD_{830} was in the range 2.3 to 2.6, indicating a high degree of purity in the preparation. Aliquots of the reaction center fraction were dialyzed for at least 24 h against 0.3 1 of 0.1 M sodium phosphate buffer containing 0.01% sodium azide, pH 5.3, at 23°C. The protein concentration of the dialysate was in the range of 1 to 2 mg reaction center protein/ml, as judged by the optical density of the reaction center fraction at 280 nm. This particular condition was optimal, although changes in ionic strength (0.01 to 0.3 M sodium phosphate) and pH (pH 5.0 to 6.6) of the dialysis fluid produced useful crystals. Probably most important was the influence of temperature on crystal formation. Initially, the recrystallization experiments were performed at 4°C with sporadic success. Changing the temperature of the dialysis to 23°C produced improved reaction center crystals.

While we tried a number of conditions to produce two-dimensional crystals, varying parameters such as temperature, protein concentration, pH, salt concentration, detergent, and purity of the sample, we were not able to use this experience to develop a general set of guidelines for the crystallization of membrane proteins. For better or for worse, the formation of such crystalline sheets is highly dependent on the nature of the individual specimen, and we have not been able to form crystals in other systems without altering some aspects of our preparation conditions.

Detergent dialysis of purified reaction centers results in the formation of regular sheets as shown in Figure 1. The sheets show a high degree of order and close packing of individual subunits. In contrast to the native *Rhodopseudomonas viridis* photosynthetic membrane, the subunits arranged within these sheets are ordered in a rectangular lattice rather than a hexagonal one.[1] Optical diffraction of images of these sheets in negative stain shows a rectangular pattern with unit cell dimensions of 121 × 129 Å. The best patterns extend to six orders in both dimensions, indicating a nominal image resolution of approximately 20 Å. The individual subunits within the lattice are clearly seen as electron-transparent, stain-excluding structures at intervals of 65 Å in either direction.

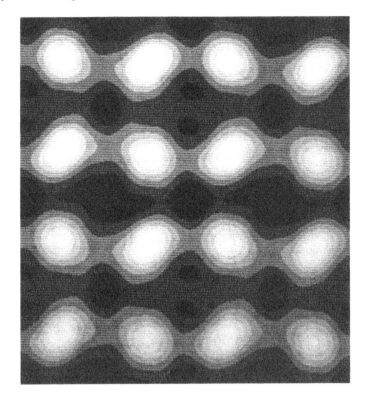

FIGURE 2.   Fourier-processed image of the reaction center crystals in negative stain. Individual stain-excluding subunits (16 are visible in this image) correspond to individual reaction centers. The total dimensions of the figure are approximately 240 Å on each side. Four unit cells are shown, each of which contains four proteins, with a twofold symmetry axis near the center of the unit cell.

Figure 2 illustrates the basic structure of the membrane sheet after Fourier enhancement of the repeating features of the regular lattice. This analysis of the reaction center shows that the basic arrangements of subunits within the crystalline sheet is rectangular. The crystallographic unit cell contains four subunits (Figure 2), each with approximate dimensions of 45 × 60 Å. These basic images allowed the first direct observation of a photosynthetic reaction center, and enabled us to assess the general size and shape of the reaction center molecule in projection. Preliminary calculations of the volume which might be occupied by a single subunit were carried out, and suggested that each subunit visible within the unit cell did indeed correspond to a single reaction center protein. The roughly elliptical shape of the reaction center visualized in these studies was confirmed by the first X-ray reports of sections through the complex,[4] validating the usefulness of this procedure in the low resolution determination of membrane protein structure.

IV. INSERTION OF THE REACTION CENTER INTO LIPOSOMES

One of the first questions which we tried to answer about the reaction center was its disposition within the photosynthetic membrane. An earlier report from our lab on the effects of proteolysis on the *Rhodopseudomonas viridis* photosynthetic membrane had shown that the four reaction center polypeptides were all accessible to rapid proteolytic attack[8] implying that each of these polypeptides are exposed on the surfaces of the membrane. This impression was confirmed by a series of simple experiments in which purified reaction centers were

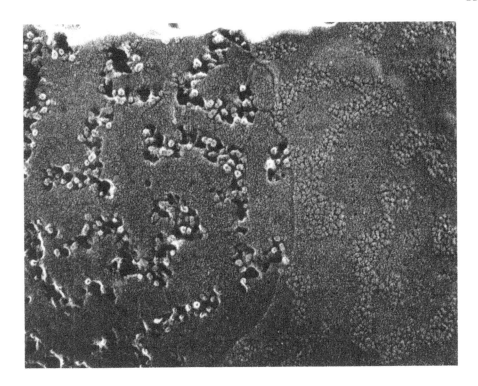

FIGURE 3. Purified *Rhodopseudomonas viridis* reaction centers can be inserted into liposomes by the freeze-thaw technique.[9] This reaction center-containing liposome has been freeze-fractured and then etched to expose the outer surface of its membrane. While the reaction centers clump to form large particles in the fracture face, the true outer surface shown at the right side of the micrograph shows particles which match the proper dimensions of the reaction center (40 to 50 Å). Their visibility at the liposomes surface indicates that these proteins span the membrane and extend well into the solvent region surrounding the lipid bilayer, an observation that has now been confirmed by other structural studies. Magnification × 140,000.

added to liposomes and examined by freeze-fracturing and deep-etching.[9] As shown in Figure 3, fractured and etched regions of the liposomes display different structures. Specifically, particles on the fractured regions are much larger (approximately 120 Å in diameter) than those on the etched regions at the surface of the liposome (45 Å).

The difference in particle diameter could be explained by distortions in the fracturing process: due to plastic deformation, aggregated groups of reaction centers fracture together to form a single large particle. However, the dimensions of particles on the etched liposome surface closely match those of the individual reaction centers from the two-dimensional crystal, indicating that individual reaction centers do appear as single structures at the surface. This observation suggested that the overall conformation of the reaction center protein enabled it to span the liposome bilayer and extend substantially from either surface into the solvent surrounding the liposome. Assuming that this would also be true for the native photosynthetic membrane, it was clear that the shape of the reaction center allowed it to extend substantially from the surface of a lipid bilayer.

V. THE STRUCTURE OF THE REACTION CENTER IN SHADOWED PREPARATIONS

In a subsequent study we then examined quick-frozen and deep-etched preparation of the crystals as shown in Figure 4.[10] The surface of the crystal differs fundamentally from the negatively stained picture. Paired rows of wedge-shaped particles are visible on the

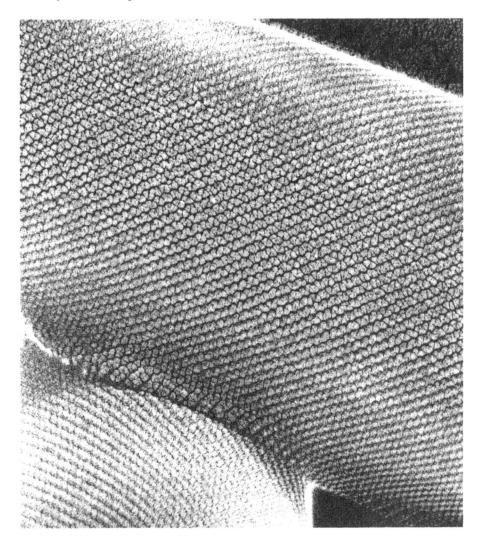

FIGURE 4.   The crystalline sheets may also be examined by quick-freezing and deep-etching. This rotary-shadowed micrograph shows the outer surface of a 2D crystal. Note the interdigitating pattern of wedge-shaped subunits visible on the surface of the shadowed crystal. This view of the crystal is quite different from the unit cell revealed in negatively stained preparations (Figures 1 and 2). Magnification × 200,000.

surface of the crystal. These rows are separated by narrow (ca. 20 Å) furrows running in parallel across the surface. The discrepancies between the freeze-etch image and the negatively stained image shown in Figure 1 and 2 should be obvious. A single crystallographic unit cell encompasses four subunits within the filtered image but only two wedge-shaped particles within the freeze-etched image. At first we found this observation puzzling.

Therefore, we searched for membrane regions where the shadowing angle seemed to be higher from some directions than others. In Figure 5 this variation of shadowing angle is clearly shown. Individual unit cells from several regions (Figure 5a) have been enlarged and diagrammed for clarity in Figures 5b to 5e.

The unit cell illustrated in Figure 5b has been shadowed from a relatively high angle in one direction. This shadowing has allowed all four subunits within the unit cell to be visualized in the replica. These four subunits followed the staggered arrangement of the negatively stained images. However, two of the four subunits have accumulated more of

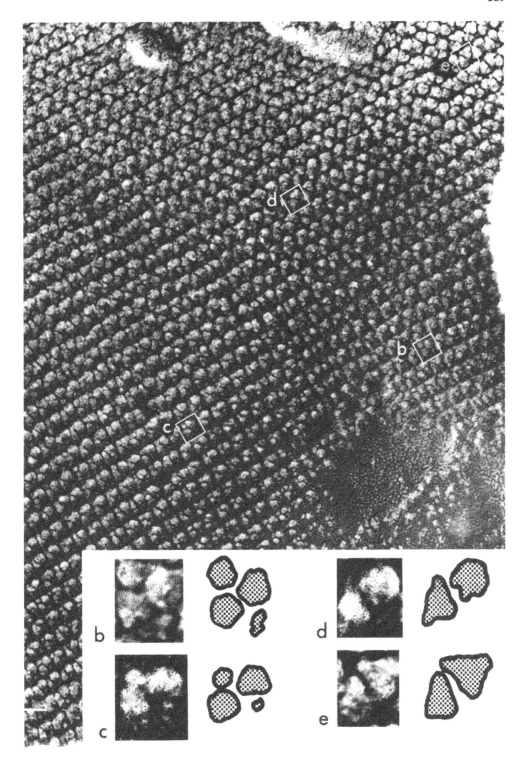

FIGURE 5. This large rotary-shadowed region of the 2D crystal helps to explain the appearance of the crystal in such preparations. Four different unit cells are marked in Figure 5a for further examination in the smaller figures: 5b through 5e. Note that in the areas with higher shadowing angles as many as four subunits are visible within the unit cell (Figures 5b and 5c), while in others only two wedge-shaped subunits can be seen (5d and 5e). This suggests an arrangement of proteins in the unit cell in which two diagonally opposite proteins are elevated with respect to the two other subunits. Magnification × 360,000 (a); 1,300,000 (b through e).

the metal shadow than two others. This suggests that the two heavily shadowed subunits may project above the surface of the sheet farther than the lightly shadowed subunits. In addition, the two heavily shadowed subunits are found at diagonally opposite corners of the unit cell.

Another unit cell, shown in Figure 5c, has been shadowed at a somewhat lower angle, and the appearance of subunits has changed. The two subunits which occupy the same position as the heavily shadowed subunits in Figure 5b are now even more pronounced. The subunits which correspond to the lightly shadowed subunits of Figure 5b are now even less visible. This further supports the suggestion that two of the subunits project farther from the crystal surface than the others.

Figures 5d and e show unit cells which have been shadowed almost uniformly at the 21° angle used to prepare our replicas. Under these circumstances, the unit cell appears identical to the typical freeze-etch crystal surface shown in Figure 4. Only two, large, wedge-shaped subunits are visible within the unit cell and occupy the same positions which are occupied in Figures 5b and c by the heavily shadowed subunits of the unit cell.

The simplest explanation of the shadowed images is that the four subunits are placed in identical positions in the crystalline sheet. Among any four subunits, two project farther from the surface of the crystal, and it is these two which accumulate the largest amount of heavy metal shadow. By accumulating the largest amount of metal during low-angle rotary shadowing, they help to form an image of the sheet which differs markedly from that obtained from negative staining.

In addition, the rotary shadowing process itself provides an explanation for the wedge-shaped particle. As illustrated in Figure 6, during the rotary shadowing process, a layer of metal is built up around the edges of a protruding particle. This coating will be uniform and round if the particle is evenly shadowed and stands alone. However, placing an adjacent particle next to it changes the nature of the shadow. Because some of the metal will be ''shaded'' by the closely placed particle, the coat of metal around the particle will be uneven. Two closely placed particles will pick up less metal during the shadowing process at the region where the two particles are closest (Figure 6b).

If this analysis is extended to include the actual arrangement of subunits within the reaction center crystals, the origin of the wedge-shaped particles becomes clear. The staggered arrangement of subunits, combined with the shading effect, produces a series of wedge-shaped particles arranged in alternating directions. It is precisely this arrangement that is observed in freeze-etch crystal preparations. Therefore, the wedge-shaped particles are consequences of the placement of elevated subunits and the rotary shadowing process itself (Figure 7). A combination of rotary shadowing with the staggered arrangement of subunits at one surface of the crystal therefore produces the wedge-shaped structure in shadowed material.

Figure 8 illustrates the implications of this data for the arrangement of protein subunits in the two-dimensional crystal. Within a single line of subunits, each protein molecule is inverted with respect to the adjacent molecules. This arrangement of subunits adds an additional symmetry axis to the crystals, making the plane group of the crystal $P22_12_1$. Therefore, a combination of structural approaches, including negative staining and freeze-etching, allowed us to determine an additional aspect of the crystal structure.

VI. A THREE-DIMENSIONAL STRUCTURE FOR THE REACTION CENTER BY ELECTRON MICROSCOPY

The highly ordered nature of the two-dimensional crystalline sheets also allowed us to carry out a determination of the three-dimensional structure of the reaction center proteins within the sheets. This direct determination of structure could then be compared to information gained from other sources, including the liposome experiments and metal shadowing

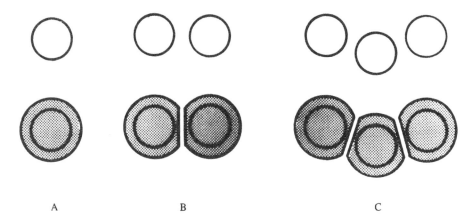

FIGURE 6. The accumulation of rotary-shadowed metal on a particle. (A) A typical particle accumulates shadow from all directions. (B) Two particles placed next to each other shade the area between them, so that each accumulates less shadow in the region where they face each other. (C) Particles arranged in staggered rows shade each other in a way that produces a wedge-shaped pattern, similar to what is seen in shadowed preparations of the raction center crystal.

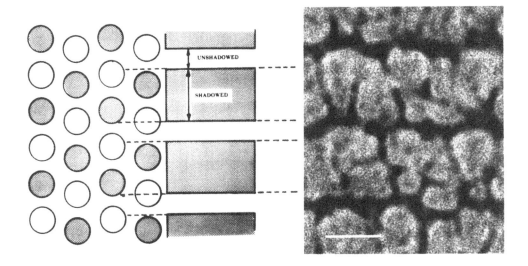

FIGURE 7. Diagram illustrating how a staggered placement of particles in which two particles from each unit cell are elevated (accumulating more shadow) can produce the wedge-shaped pattern seen in the 2D crystals. An enlarged portion of the crystal surface is shown for comparison. Bar = 0.01 μm.

results which have just been mentioned. The data for the three-dimensional analysis was obtained from a set of 37 electron micrographs of reaction center two-dimensional crystals, contrasted with 2% uranyl acetate.[11] The set of micrographs was collected at tilt angles up to 68°, providing a set of Fourier transforms from which amplitude and phase data could be sampled along 31 different independent lattice lines.

An estimate of the 0,0 lattice line in three dimensions cannot be obtained from tilted specimens, and edge-on views of the reaction center sheets were used to determine the values for this portion of the three-dimensional transform. Amplitude and phase terms from each micrograph in the data set were combined by means of an image-combining program. Images from the data set showed an average phase residual of 24° during the combining process. Representative phase and amplitude data from the lattice lines were interpolated to assemble the data used for the reverse three-dimensional transform. The data included terms

FIGURE 8.   The arrangement of one row of subunits in the crystalline sheet, as predicted on the basis of shadowing data.[10] The head-to-tail arrangement predicted in this image was verified by a final three-dimensional map of the reaction center in negative stain.

to the 6th order, indicating a resolution of approximately 20 Å. Individual micrographs of untilted specimens displayed clear P2 symmetry, and the three-dimensional analysis was carried out with this symmetry applied to the data.

The process of three-dimensional reconstruction by Fourier techniques allows us to determine the arrangement of these four subunits within the unit cell. Such a reconstruction gives an accurate image of the actual structure of each subunit, subject to the basic limitations of the technique, namely, the fact that only stain-excluding regions are visible, the specimen is subject to distortion and radiation damage in the electron microscope, and that resolution is limited by the very process of dehydration and staining to about 20 Å in these specimens.

The reconstructed three-dimensional data have been sampled by drawing contours at fixed intervals in order to build up a three-dimensional map of the structure. In Figure 9 we have taken this approach by selecting a single contour value as the edge of the structure and displaying slices of the structure at regular intervals on a computer graphics device.[11] This technique produces clear images of each of the four subunits within a crystallographic unit cell. As shown in Figure 9, each subunit is an extended ellipsoid, approximately 50 Å in diameter at its thickest point, and extending from one edge of the crystalline sheet to the other. Each subunit is inclined 20° from the axis normal to the sheet. The length of each subunit in the unit cell is approximately 85 Å.

A rough estimate of the molecular mass of the *Rhodopseudomonas viridis* reaction center protein is 135,000 daltons (Da) (obtained by adding the published values for the H, M, and L subunits of the reaction center[12,13] to an estimate of 41,000 Da for cytochrome). If one assumes a density of 1.3 Å³ per Da, then an aggregate molecular volume for the complex of approximately 175,000 Å³ can be predicted, neglecting the contribution of bound pigments and prosthetic groups to the overall volume. A rectangular solid measuring $50 \times 35 \times 85$ Å has a volume of 148,750 Å³. It is true that uncertainties abound in estimates made of structures embedded in negative stain, nonetheless, a reduction of molecular volume of approximately 15% is to be expected due to the drying effects of the staining process. Given these figures, it is clear that the size of a single subunit in the unit cell is comparable to what we might expect from a single photosynthetic reaction center.

The extended shape of the reaction center makes it clear how this structure might extend from both surfaces of the photosynthetic membrane. It also implies that there is but one way to incorporate the reaction center map into the three-dimensional map of the *Rhodopseudomonas viridis* membrane itself: as the large central mass which spans the membrane.[2] This conclusion would not be possible except for the unique nature of the *Rhodopseudomonas viridis* system, in which a biological membrane and one of its components can both be analyzed by Fourier techniques.

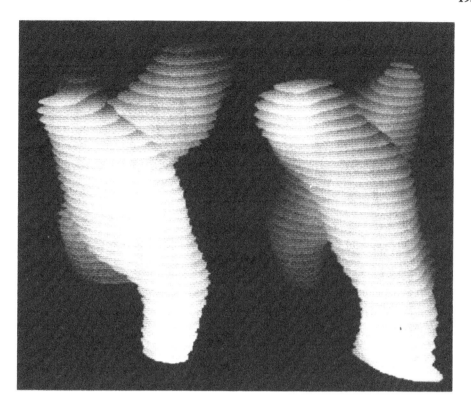

FIGURE 9. A three-dimensional map of the reaction center crystal was determined by Fourier image analysis from tilted images of the crystal in negative stain. This graphic representation of a single unit cell shows four reaction center proteins. Each reaction center is roughly elliptical in cross-section, and extends well past either surface of the biological membrane: its maximum contour length is 85 Å. The overall size and shape of the reconstruction is similar to the map determined by X-ray data,[4-6] although the dimensional of the protein are considerably reduced, probably due to the drying process associated with negative stain.

VII. RELATIONSHIP OF THE EM IMAGE TO X-RAY DATA FROM THE REACTION CENTER

Michel's three-dimensional crystals of the *Rhodopseudomonas viridis* reaction center[3] have enabled a three-dimensional map to be produced at the atomic level of resolution,[4-6] and have yielded interesting information concerning the organization of prosthetic groups in the central region of the reaction center.[4] Although the EM map is necessarily at much lower resolution, its general features can be compared to the map which has been derived by electron microscopy. The general size and shape of the reaction center determined by electron microscope—an extended membrane-spanning structure with an elliptical cross-section—is in good agreement with detailed maps prepared from X-ray data. However, Deisenhofer et al.[4-5] have reported a maximum diameter for other regions of the reaction center of 70 Å, which we have not seen at any level of the EM map. They have also reported that the maximum dimension of the map is 125 Å. This value is larger than any distance through which we have been able to trace the EM map.

There are, in general terms, two ways to explain the increased length of the reaction center in the X-ray map. Specimen dehydration, coupled with heavy metal staining, may have decreased the apparent length of the reaction center in the EM map. Because these factors do not affect the X-ray data, the larger dimensions may reflect the authentic dimensions of the reaction center. While the extraordinary (3 A) detail of the X-ray map is beyond dispute, one noteworthy aspect of the *Rhodopseudomonas viridis* reaction center crystals is their very high (70%) solvent content.[3] In addition, the 223 × 223 × 114 Å unit cell dimension of the three-dimensional crystals is much larger than 121 × 129 × 85 Å dimensions of the two-dimensional crystals, suggesting that the packing density of reaction centers in the three-dimensional crystals is much less than it is in the two-dimensional sheets. It is possible that the low protein density and the high solvent content of the three-dimensional crystals may have caused the unfolding of some molecular groups in the reaction center, resulting in a map which may be somewhat larger in some dimensions than the native membrane-bound reaction center.

As noted earlier, deep-etching and rotary shadowing was used by our laboratory[10] to examine the surfaces of two dimensional crystalline sheets formed from the *Rhodopseudomonas viridis* reaction center. The results of this study suggested to us that the actual arrangement of four subunits within the unit cell corresponded to a symmetry of $P22_12_1$. Although I have assumed only P2 symmetry in calculating the three-dimensional map, I believe that the higher symmetry is in fact correct for the arrangement of subunits within the sheet. Two major features are suggested by the shadowing work: first, a staggered arrangement of subunits at each surface of the sheet, and second, that each of the four subunits should be shifted off the central plane of the sheet so that two were closer to one surface, and two closer to the other. The first of these two features is in fact observed if one examines contour maps near either the top or bottom surface of the crystal in Figure 9.

The second prediction of the shadowed work is not observed, however. This can best be explained by the EM technique used for the reconstruction. Our shadowing study used the deep-etching method in which frozen (hydrated) membrane samples are used for rotary replication. The three-dimensional reconstruction was carried out on negative-stained material, in which membrane samples are dried down against a carbon film and dehydrated. During the drying process, the compression of the sheet against the film may have squeezed the sheet together so that all four subunits were compressed into roughly the same central plane. The drying process may explain, therefore, both the deviation from $P22_12_1$ symmetry, and the reduced size of the reaction center when compared to the X-ray map.

One of the principal questions left unanswered by this study is the structural nature of the association between the light-harvesting and reaction center components of the membrane. As work continues on these and other questions, it seems likely that the *Rhodopseudomonas viridis* photosynthetic membrane will prove an invaluable system in which to explore the molecular details of photosynthesis.

ACKNOWLEDGMENTS

I am grateful for the assistance, support, and collaboration of Jules S. Jacob. This work was supported by a grant from the National Institutes of Health of the United States (GM 28799).

REFERENCES

1. **Miller, K. R.,** Structure of a bacterial photosynthetic membrane, *Proc. Natl. Acad. Sci. U.S.A.,* 76, 6415, 1979.
2. **Miller, K. R.,** Three dimensional structure of a photosynthetic membrane, *Nature,* 300, 53, 1982.

3. **Michel, H.,** Three dimensional crystals of a membrane protein complex. The reaction center from *Rhodopseudomonas viridis, J. Mol. Biol.,* 158, 567, 1982.

4. **Deisenhofer, J., Epp, O., Miki, K., Huber, R., and Michel, H.,** X-ray structure analysis of a membrane protein complex. Electron density map at 3 Å resolution and a model of the chromophores of the photosynthetic reaction center from *Rhodopseudomonas viridis, J. Mol. Biol.,* 180, 385, 1984.

5. **Deisenhofer, J., Epp, O., Miki, K., Huber, R., and Michel, H.,** Structure of the protein subunits in the photosynthetic reaction centre of *Rhodopseudomonas viridis* at 3 Å resolution, *Nature,* 318, 618, 1985.

6. **Michel, H. and Deisenhofer, J.,** The structural organization of photosynthetic reaction centers, in *Progress in Photosynthesis Research,* Vol. 1, Biggins, J., Ed., Martinus Nijhoff, Dordrecht, 353.

7. **Miller, K. R. and Jacob, J. S.,** Two dimensional crystals formed from photosynthetic reaction centers, *J. Cell Biol.,* 97, 1266, 1983.

8. **Jacob, J. S. and Miller, K. R.,** Structure of a bacterial photosynthetic membrane. Isolation, polypeptide composition, and selective proteolysis, *Arch. Biochem. Biophys.,* 223, 282, 1983.

9. **Miller, K. R. and Jacob, J. S.,** Photosynthetic reaction centers in artificial membranes: estimating protein dimensions by freeze-fracture and freeze-etching, *J. Submicrosc. Cytol.,* 16, 619, 1984.

10. **Miller, K. R. and Jacob, J. S.,** Two dimensional crystals of a membrane protein: arrangement of subunits within the crystal sheet, *Eur. J. Cell Biol.,* 36, 247, 1985.

11. **Miller, K. R.,** Structural analysis of a photosynthetic membrane: *Rhodopseudomonas viridis,* in *Advances in Cell Biology,* Vol. 1, Miller, K. R., Ed., JAI Press, New York, 1987, 131.

12. **Michel, H., Weyer, K. A., Gruenberg, H., and Lottspeich, F.,** The 'heavy' subunit of the photosynthetic reaction center from *Rhodopseudomonas viridis,* isolation of the gene, nucleotide and amino acid sequence, *EMBO J.,* 4, 1667, 1985.

13. **Michel, H., Weyer, K. A., Gruenberg, H., Dunger, I., Oesterhelt, D., and Lottspeich, F.,** The 'light' and 'medium' subunits of the photosynthetic reaction center from *Rhodopseudomonas viridis:* isolation of the genes, nucleotide and amino acid sequences, *EMBO J.,* 5, 1149, 1986.

Chapter 11

PREPARATION OF MEMBRANE CRYSTALS OF MITOCHONDRIAL NADH:UBIQUINONE REDUCTASE AND UBIQUINOL:CYTOCHROME C REDUCTASE AND STRUCTURE ANALYSIS BY ELECTRON MICROSCOPY

Hanns Weiss and Kevin R. Leonard

TABLE OF CONTENTS

I. INTRODUCTION

The mitochondrial system of oxidative phosphorylation is contained in the inner membrane in the form of four discrete enzyme complexes. They are the NADH:ubiquinone oxidoreductase (NADH:Q reductase), the ubiquinol:ferricytochrome c oxidoreductase (cytochrome reductase), the ferrocytochrome c:oxygen oxidoreductase (cytochrome oxidase), and the ATP-synthase. The three electron transfer enzymes are concerned with the generation of transmembraneous protonic potential. They link downhill electron flow with uphill proton translocation across the membrane directed outwards from the mitochondria. The ATP-synthase is the main consumer of the protonic potential and couples downhill and inward flow of protons with synthesis of ATP (for a review see Reference 1).

The two most remarkable features of the proton translocating enzymes of mitochondria are their enormous subunit complexity and the dual genetic origin of subunits. The NADH:Q reductase, for example, consists of some 25 different subunits (depending on the source of mitochondria) of which 7 are encoded by mitochondrial DNA and the remaining by nuclear DNA (for a review see Reference 2). Cytochrome reductase is composed of 9 to 12 different subunits of which 1 is mitochondrial and the remaining are nuclear-encoded (for a review see Reference 3).

The general reaction catalyzed by the NADH:Q reductase is

$$NADH + Q + 5 H^+_N \rightarrow NAD^+ + QH_2 + 4 H^+_P$$

where N and P refer to the negative inner and positive outer side of the membrane. The electron transfer is carried out with the aid of one FMN and six FeS clusters, whose sequence of operation and mechanisms of linkage to proton translocation are largely unknown. The reaction is specifically inhibited by rotenone and piericidin.[1,2]

The reaction catalyzed by cytochrome reductase is

$$QH_2 + 2 \text{ ferricytochrome c} + 2 H^+_N$$
$$Q + 2 \text{ ferrocytochrome c} + 4 H^+_P$$

Specific inhibitors are antimycin and myxothiazol, which block the electron transfer reaction independently at two different Q-catalytic centers.[1,3]

II. ISOLATION OF THE ENZYMES IN DETERGENT SOLUTION

NADH:Q reductase was isolated from *Neurospora crassa* mitochondria by fractional extraction of the membrane protein with Triton X-100, followed by chromatography on DEAE-Sepharose and size exclusion HPLC performed in solutions containing 0.05% Triton X-100.[4,5] A monodisperse preparation was obtained which consists of protein, phospholipid, and Triton at a weight ratio of 1:0.04:0.15. From sedimentation equilibrium distribution of the preparation a protein molecular weight of 650,000 resulted. This value approximates to the sum of apparent molecular weights of about 25 subunits as resolved by SDS-gel electrophoresis. When isolated, therefore, NADH:Q reductase must be in a monomeric state.[5]

Isolation of cytochrome reductase as a protein-phospholipid-Triton complex (1:0.03:0.22 by weight) was achieved by solubilization of mitochondrial membrane protein with Triton X-100, followed by affinity chromatography on cytochrome c coupled to Sepharose and subsequent gel filtration.[6] Sedimentation equilibrium and small angle neutron scattering analysis of the preparation indicated a protein molecular weight of about 550,000. The value is approximately twice the sum of the apparent molecular weights of subunits indicating that the enzyme is isolated in the dimeric state.[6-8]

TABLE 1
Conditions that Lead to the Formation of Membrane Crystals

Protein	Crystal form	Phospholipid to protein ratio	pH	NaCl (mM)	Detergent removal	Temperature (°C)	Ref.
NADH:Q reductase	Monolayer	0.3—0.4	8.0	200	Dialysis	20	5
Cytochrome reductase	Monolayer	2.0—0.5	7.0	0	Adsorption	4	10
	Multilayer	2.0—0.5	5.5	50	Dialysis	20	12
bc₁-Subcomplex	Monolayer	2.0—0.5	5.5	50	Dialysis	20	11

By treatment with NaCl (>0.2 M) in Triton solution the cytochrome reductase was dissociated into three parts: first, a so-called bc_1-subcomplex which consists of the subunits III (cytochrome b), IV (cytochrome c_1), and VI to IX (four small subunits of unknown function), second, a subcomplex of the two large subunits I and II (the so-called core-complex), and, third, the single subunit V (the so-called Rieske-FeS-protein). The preparation of the bc_1-subcomplex was separated from the other two parts by gel filtration in Triton solution. It consists of protein, phospholipid and Triton (1:0.01:0.4 by weight), has a protein molecular weight of about 280,000 and is in dimeric state like the whole enzyme.[8,9]

III. PREPARATION OF MEMBRANE CRYSTALS

The growth of membrane crystals was systematically attempted with the NADH:Q reductase, the cytochrome reductase, and the bc_1-subcomplex. Only preparations that eluted from the gel filtration or size exclusion HPLC columns as symmetric peaks of protein concentrations above 2 mg/ml were used. They were incorporated into phospholipid bilayers by incubating the protein with Triton-phospholipid micelles and then removing the Triton.

For preparing the phospholipid-Triton micelles, 20 mg of the phospholipid was dissolved in 2 ml of chloroform, dried under a stream of nitrogen, dissolved in 2 ml of diethyl ether and dried again. The lipid was sonicated for 2 to 3 min in 4 ml of a solution containing 0.5% Triton X-100 and the buffer used for the last purification step of the protein until a homogeneous solution was obtained. The following parameters were varied: protein to phospholipid ratio, type of phospholipid, method of detergent removal, temperature, pH value, and composition of the buffer. The conditions that lead to the formation of membrane crystals are summarized in Table 1.

Crystals of NADH:Q reductase grew to a sufficient size only at a low protein to phospholipid ratio of 0.3 to 0.4. In contrast, the lipid to protein ratio was not critical for crystallization of cytochrome reductase and the bc_1-subcomplex as long as it was in the range of 0.5 to 2. It appears then that phase separation takes place such that in addition to the membrane crystals that are lipid bilayers saturated with protein there are protein-free lipid bilayers if lipid is in excess.

A. MEMBRANE CRYSTALS OF NADH:Q REDUCTASE

Samples of 50 µl of 2 to 3 mg/ml NADH:Q reductase in 50 mM Tris-acetate pH 7, 0.1 M NaCl and 0.05% Triton X-100 were mixed with 5 to 10 µl of 0.5% Soybean phosphatidylcholine (Sigma, IIS) in the same buffer solution but with 0.5% Triton and dialyzed in 2-mm tubes (Servopor, Serva) against 15 ml of buffers containing 50 mM Tris-acetate pH 8.0 and 0.2 M NaCl for 2 days at 20°C. The crystals could be stored at 4°C for a few days.[5]

B. SINGLE-LAYER MEMBRANE CRYSTALS OF CYTOCHROME REDUCTASE

Routinely, 0.4% phosphatidylcholine from Soybean (Sigma, IIS) and 0.1% phosphatidylserine from bovine brain (Roth) were dissolved in 0.5% Triton X-100, 50 mM Tris-acetate pH 7, 1 mM EDTA. Equal volumes of 5 mg/ml enzyme solution and the phospholipid-Triton solution were mixed. Triton was removed by gently stirring 0.25 g activated Bio-Beads SM-2 (Bio-Rad) per milliliter in the mixture for 2 h. The membrane suspension obtained was separated from the beads by filtration through loosely packed glass wool. Large membranes could be separated from smaller ones and from excess phospholipid by gel filtration [0.2 ml samples with a 0.5 × 7 cm Sepharose 2B (Pharmacia) column] in 50 mM Tris-acetate pH 7. Fractions of two drops were collected. After addition of 10% (w/v) sucrose the membrane crystals could be stored in liquid nitrogen temperature, for months.[10]

C. MULTILAYER MEMBRANE CRYSTALS OF CYTOCHROME REDUCTASE

In our crystallization protocol, Soybean phosphatidylcholine (Sigma, IIS) was used as lipid. Equal aliquots of enzyme and lipid solutions were combined and the detergent removed by dialyzing for 2 days at room temperature against a buffer containing 50 mM MES (morpholinoethane sulfonate) pH 5.5, 50 mM NaCl, 1 mM EDTA, and 1 mM ascorbate.[12]

D. MEMBRANE CRYSTALS OF THE bc$_1$-SUBCOMPLEX

The subcomplex obtained from the gel filtration step was concentrated by ultrafiltration (Diaflo PM 30 filter) to a protein concentration of 4 mg/ml. An equal amount of a solution of 0.4% Soybean phosphatidylcholine and 0.1% bovine brain phosphatidylserine in 0.5% Triton X-100, 50 mM Tris-acetate pH 7, 50 mM NaCl, 1 mM EDTA and 1 mM ascorbate was added. Triton was removed by dialysis for 2 days at room temperature against 50 mM MES, pH 5.5, 50 mM NaCl, 1 mM EDTA and 1 mM ascorbate.[11]

IV. STRUCTURE ANALYSIS BY ELECTRON MICROSCOPY

Specimens on carbon-covered collodion films were stained with 1% (w/v) uranyl acetate. Electron microscopy was carried out at 80 kV using a Philips 400 T instrument with 60° eucentric tilt specimen holder. Micrographs taken at 34,000× magnification were digitized with an Optronics drum densitometer at 25 μm raster corresponding to a pixel size of 7.35 Å. Data for four sets of tilted images were combined to give a three-dimensional reconstruction.[5,7,9]

A. NADH:Q REDUCTASE

Membrane crystals were obtained as vesicles, tubes or sheets (Figure 1). At the edges of tubes where the membrane folds over protein molecules could be seen projecting about 10 nm at regular intervals along the edges of the crystal. Diffraction patterns of untilted crystals extended to 3.9 nm with a unit cell size of 19 nm × 38 nm, γ = 90°, with systematic absences along the principal axes for odd values of h and k (Figure 2). The two-sided planed group packing corresponding to projection symmetry of pgg is p22$_1$2$_1$. In this symmetry the alternate enzyme molecules point up and down across the membrane and are related by a twofold screw axis parallel to the plane of the membrane.

The result of the three-dimensional reconstruction is shown in Figure 3. The density map has been contoured at a level where there is contact between molecules within the rows which run along the unit cell a-axis direction but the weak density between the rows is lost. The volume of the stain-excluding density at this contour level is about 2.5 × 10^6 Å.^{3} Since there are four molecules per unit cell, this corresponds to molecular weight (M$_r$) of 500,000 per monomer which is about 80% of the M$_r$ value obtained from sedimentation equilibrium distribution of the protein in detergent solution.

In this reconstruction, as was the case for cytochrome reductase and the bc$_1$-subcomplex (see below), there is no clear boundary for the membrane. The plot of contrast difference

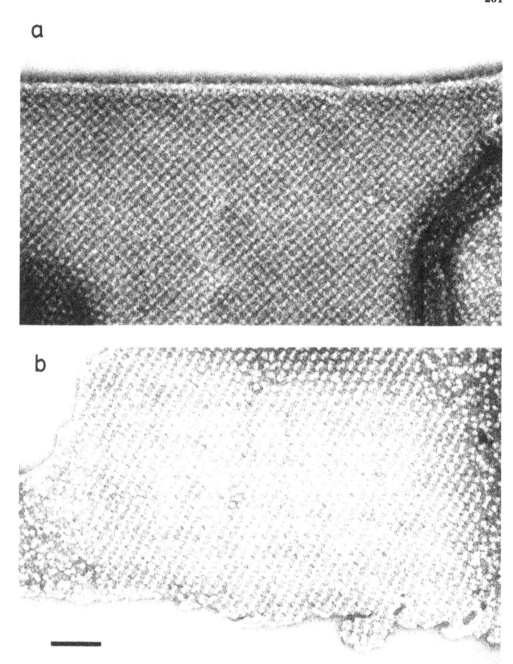

FIGURE 1. Membrane crystals of NADH:Q reductase. (a) Edge of a tubular membrane crystal. The complex moiré pattern arises from the overlap of the two layers. At the edge, where the crystals fold over, molecules can be seen in cross-section projecting about 10 nm from the surface. (b) Single layer membrane crystal showing rows of ordered protein molecules separated by about 20 nm. At the left and right of the image, membrane regions without crystallinity can be seen with single protein molecules arranged randomly in the bilayer. The scale bar represents 100 nm.

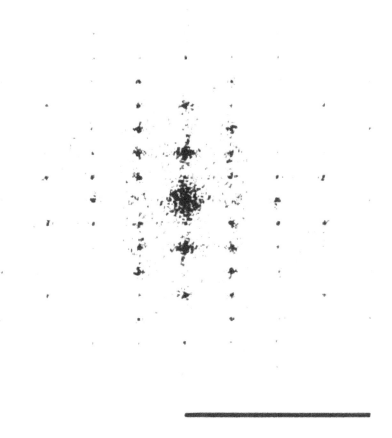

FIGURE 2. Computer generated Fourier transform of the image shown in Figure 1a. The diffraction spots extend to about 3.9 nm in reciprocal space, with systematic absences along the a- and b-axes and fairly good mirror symmetry indicating plane group pgg. The scale bar represents a reciprocal distance of 0.21 nm^{-1}.

vs. the c-direction of the unit cell did not show a strong decrease towards the center of the structure where the bilayer is present. Since the alternating enzymes are arranged in an up-and-down manner across the membrane, a 5-nm thick layer in the middle of the cross-section through the unit cell in c-direction was taken to be the position of the bilayer (Figure 3B).

In order to divide the stain excluding density of the crystal to a single NADH:Q reductase molecule, the map was separated at the narrow regions between the large projecting domains. This gave a particle which is clearly asymmetric with respect to the phospholipid bilayer. About two thirds of the total protein projects from one side of the bilayer (Figure 6).

The electron microscopic studies gave no evidence on the sideness of this large peripheral part with respect to the mitochondrial inner membrane. Degradation of bovine heart NADH:Q reductase permitted the isolation of a water-soluble flavoprotein fragment consisting of three polypeptides (M_r 51,000, 24,000 and 10,000) in a 1:1:1 molar proportion. By photoaffinity labeling the NADH-binding site was localized to the 51,000 M_r polypeptide.[2] Since the small peripheral part of the structure cannot account for the protein mass of the flavoprotein fragment, it is reasonable to suggest that the large peripheral part of NADH:Q reductase projects into the matrix space of mitochondria (Figure 6).

B. CYTOCHROME REDUCTASE AND bc₁-SUBCOMPLEX

Single layer sheets, vesicles or tubes were obtained from cytochrome reductase by the detergent adsorption method at pH 7.0 and 4°C (Figure 4a). The diffraction pattern of these crystals extend to the fifth order which corresponds to a resolution of 2.5 nm.[7,12]

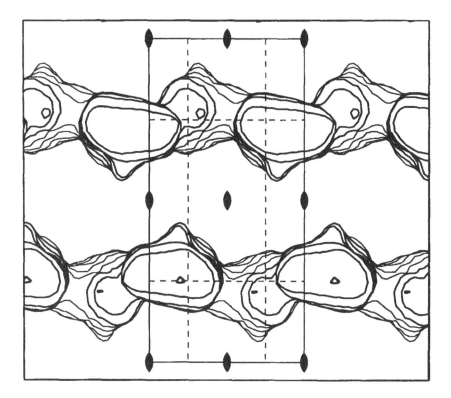

A

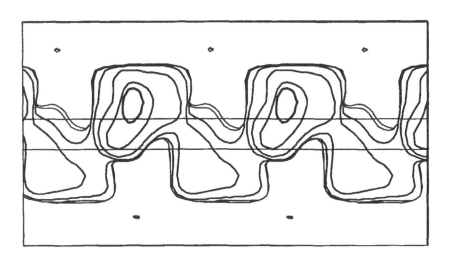

B

FIGURE 3. Contour maps of NADH:Q reductase. (A) A hidden line contour map of the three-dimensional reconstruction seen in the view normal to the plane of the membrane. The two-sided plane group is p22$_1$2$_1$ with large domains of density projecting alternately above and below the plane of the page. The dimensions of the unit cell are 19 nm × 38 nm. (B) Hidden line presentation of the map projecting along the unit cell b-axis. The central parallel lines indicate the position of the bilayer. The large domains of density that persumably corresponds to individual molecules are arranged so that most of the stain-excluding density would be on one side of the bilayer.

Multilayer membrane crystals of cytochrome reductase were formed by the dialysis method at pH 5.5 and room temperature. The membrane crystals stack in register on top of each other to give three-dimensional crystals several layers thick (Figure 4b). Diffraction patterns extend to the tenth order, which corresponds to 1.8 nm resolution.[12]

Membrane crystals of the bc_1-subcomplex were obtained only by the dialysis method at pH below 6.0 and at room temperature. Monolayer tubes vesicles and sheets typically were formed. The diffraction pattern extend to the fifth order.[11]

The dimensions of the rectangular unit cells of the monolayer crystal were 13.7 nm $\times$ 17.4 nm with the whole enzyme, and 9.4 nm $\times$ 14.3 nm with the bc_1-subcomplex. The symmetry of both crystals was $p22_12_1$ indicating that the alternating (dimeric) protein molecules point up and down across the membrane. In contouring the maps of the three-dimensional reconstruction, cut off levels were chosen so that volumes of the calculated structure corresponded to about 80% of the volumes expected for the molecular weights of the proteins (Figures 5 and 6).

The structures show that the enzyme and the bc_1-subcomplex are dimers with monomeric units related by a twofold axis running perpendicular to the membrane. In the structure of the enzyme three regions can be distinguished, a large peripheral section which protrudes 7 nm from the one side of the membrane and accounts for 50% of the total protein, a membrane section which contains 30% of the protein and a small peripheral section which protrudes 3 nm from the opposite side of the membrane and accounts for 20% of the protein. The two enzyme monomers contact each other partly in the membrane and partly in the large peripheral section. The structure of the bc_1-subcomplex shows two peripheral lobes which correspond in size and shape to the small peripheral section of the enzyme. Towards the membrane the protein merges into one less defined region of about 4 nm in thickness and roughly corresponds to the membrane section of the enzyme. The structure shows no part large enough to fit to the large peripheral section of the enzyme.

On the basis of the three-dimensional structures in combination with the folding of subunits predicted from primary structures[14-16] and the solubility properties of isolated subunits or subunit domains[8,17,18] the assignment of subunits to either of the two peripheral sections of the enzyme or the membrane section was made: The isolated subcomplex of the subunits I and II (the core-proteins) is water-soluble and must therefore lie peripherally with regard to the membrane.[8] The molecular weight of this subcomplex accounts well for the size of the large peripheral section in the structure of the enzyme. Cytochrome b is predicted to span the membrane with 8 to 9 hydrophobic segments and must therefore lie predominantly in the membrane.[14] Cytochrome c_1[15,17] and the iron-sulfur subunit[16,18] are amphiphilic proteins each consisting of a large hydrophilic part to which the redox group is attached and a smaller hydrophobic segment which anchors the subunit to the membrane. The catalytic domains of these subunits must therefore lie in the small peripheral section of the enzyme.

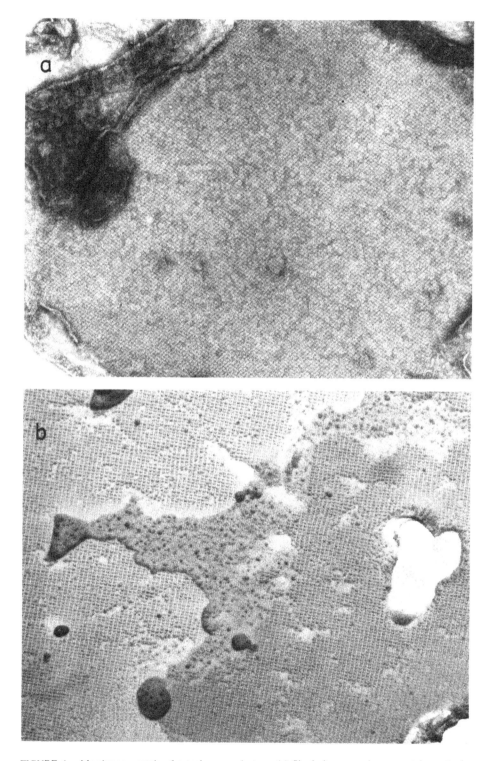

FIGURE 4. Membrane crystals of cytochrome reductase. (a) Single layer membrane crystal negatively stained with uranyl acetate. (b) Multilayer membrane crystal, first negatively stained, dried, and then shadowed with platinum.

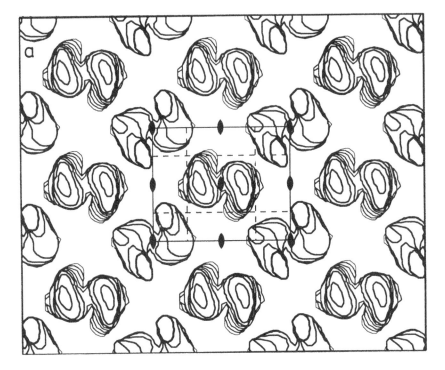

A

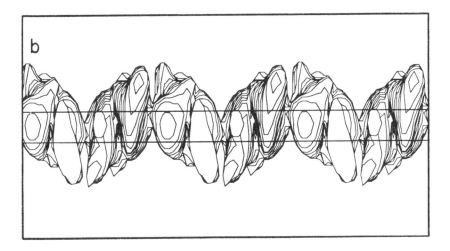

B

FIGURE 5. Contour maps of cytochrome reductase. (A) The c projection with the unit cell 13 nm × 17 nm, which shows the dimeric molecules arranged on the crystallographic dyad axes. (B) The b projection which shows the distribution of the stain excluding density across the bilayer. The scale is the same as that of NADH:Q reductase shown in Figure 3.

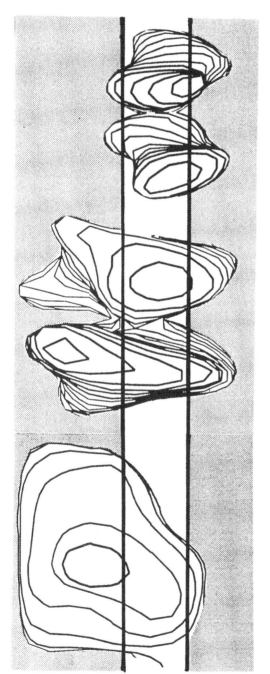

FIGURE 6. Projections in b-direction of NADH:Q reductase (left), cytochrome reductase (middle) and the bc_1-subcomplex (right). The white horizontal section represents the bilayer, the upper dark area the matrix space of the mitochondria and the lower dark area the intermembrane space.

REFERENCES

1. **Hatefi, J.,** The mitochondrial electron transport and oxidative phosphorylation system, *Annu. Rev. Biochem.,* 54, 1015, 1985.
2. **Ragan, C. I.,** Structure of NADH:ubiquinone reductase (complex I), in *Current Topics in Bioenergetics,* Vol. 15, Lee, C. P., Ed., Academic Press, London, 1987, 1.
3. **Weiss, H.,** Structure of mitochondrial ubiquinol:cytochrome c reductase (complex III) in *Current Topics in Bioenergetics,* Vol. 15, Lee, C. P., Ed., Academic Press, London, 1987, 67.
4. **Ise, W., Haiker, H., and Weiss, H.,** Mitochondrial translation of subunits of the rotenone sensitive NADH:ubiquinone reductase in Neurospora mitochondria, *EMBO J.,* 4, 2075, 1985.
5. **Leonard, K., Haiker, H., and Weiss, H.,** Three dimensional structure of NADH:ubiquinone reductase (complex I) from Neurospora mitochondria determined by electron microscopy of membrane crystals, *J. Mol. Biol.,* 194, 277, 1987.
6. **Weiss, H. and Kolb, H. J.,** Isolation of mitochondrial succinate:ubiquinol reductase, cytochrome c reductase and cytochrome c oxidase from Neurospora using nonionic detergent, *Eur. J. Biochem.,* 99, 139, 1979.
7. **Leonard, K., Wingfield, P., Arad, T., and Weiss, H.,** Three dimensional structure of ubiquinol:cytochrome c reductase from Neurospora mitochondria determined by electron microscopy of membrane crystals, *J. Mol. Biol.,* 149, 259, 1981.
8. **Perkins, S. J. and Weiss, H.,** Low resolution structural studies of mitochondrial ubiquinol:cytochrome c reductase in detergent solution by neutron scattering, *J. Mol. Biol.,* 168, 847, 1983.
9. **Karlsson, B., Hovmöller, S., Weiss, H., and Leonard, K.,** Structural studies of cytochrome reductase. Subunit topography determined by electron microscopy of membrane crystals of a subcomplex, *J. Mol. Biol.,* 165, 287, 1983.
10. **Wingfield, P., Arad, T., Leonard, K., and Weiss, H.,** Membrane crystals of ubiquinoe:cytochrome c reductase from Neurospora mitochondria, *Nature,* 280, 696, 1979.
11. **Hovmöller, S., Leonard, K., and Weiss, H.,** Membrane crystals of a subcomplex of mitochondrial cytochrome reductase containing the cytochromes b and c_1, *FEBS Lett.,* 123, 118, 1981.
12. **Hovmöller, S., Slaughter, M., Berriman, J., Karlsson, B., Weiss, H., and Leonard, K.,** Structural studies of cytochrome reductase. Improved membrane crystals of the enzyme complex and crystallization of a subcomplex, *J. Mol. Biol.,* 165, 401, 1983.
14. **Saraste, M.,** Location of the heme binding sites in the mitochondrial cytochrome b, *FEBS Lett.,* 106, 367, 1984.
15. **Römisch, J., Tropschug, M., Sebald, W., and Weiss, H.,** The primary structure of cytochrome c_1 from *Neurospora crassa, Eur. J. Biochem.,* 164, 111, 1987.
16. **Harnisch, U., Weiss, H., and Sebald, W.,** The primary structure of the iron-sulfur-subunit of ubiquinol-cytochrome c reductase from Neurospora determined by cDNA and gene sequencing, *Eur. J. Biochem.,* 149, 95, 1985.
17. **Li, Y., Leonard, K., and Weiss, H.,** Membrane bound and water soluble cytochrome c_1 from Neurospora mitochondria, *Eur. J. Biochem.,* 116, 199, 1981.
18. **Li, Y., De Vries, S., Leonard, K., and Weiss, H.,** Topography of the iron-sulfur subunit in the mitochondrial ubiquinol:cytochrome c reductase, *FEBS Lett.,* 135, 277, 1981.

APPENDIX I

Properties of Commonly Used Nonionic and Zwitterionic Detergents for Membrane Protein Solubilization and Crystallization (Compiled by M. Zulauf and H. Michel)

Generic name (short names, trade names)	Structural formula	Critical micelle concentration[a] (in H₂O)		Molecular weight	Suppliers[c]
		mM	%(w/w)		
Polyethylenoxide (PEO) Headgroups					
n-Alkyl PEO monoether	C_mH_{2m+1}-E_n-OH ($E = OCH_2CH_2$)	m = 8, n = 5	6.0	0.21	350	B, N
		10, 6	0.90	0.038	422	F, Si
((CₘEₙ; Brij-Series;		12, 8	0.071	0.0038	538	
LubrolPX		14, 8	0.009*	0.0005*	566	
(m = 12,14; n = 9.5)		16, 8	0.0021*	0.00012*	594	
LubrolWX						
(m = 16,18; n = 16.4))						
n-Alkyl phenyl PEO	C_mH_{2m+1}-Φ-E_n-OH (Φ = phenyl)	12, 9	0.1*	0.0058*	582	Si, Se
		Triton N57(m = 8)		0.018		F,
		Triton N101(m = 9)		0.018		
Tetramethylbutylphenyl PEO (Triton X, Nonidet P)	TMB-Φ-E_n-OH (TMB = 1,1,3,3 tetramethylbutyl)	n = 4.5 (Triton X45)	0.11*	0.0044*	404	F, Se
		n = 7,8 (Triton X114)	0.20*	0.028*	537	Si, C
		n = 9,10 (Triton X100)	0.24*	0.021*	625	
		n = 40 (Triton X405)	0.81*	0.159*	1966	
PEO polysorbates (Tween)	$C_mH_{2m+1}CO-E_w-CH_2-CH_2-CH-E_y-OH$ (W+X+Y+Z=20)	m = 12 (Tween 20)		0.08	1228	F, Si
		16 (Tween 40)		0.012		F, Si
		18 (Tween 60)		0.0027		
		9 = 9 (Tween 80)		0.10	1310	
Sugar Derivatives as Headgroups					
Alkyl-β-D-glucopyranoside (glucoside)		m = 6	250	6.6	264	C, Si, Se
		7	79	2.2	278	
		8	30.3*	0.89	292	
		9	6.5	0.20*	306	
		10	2.6	0.08	320	

APPENDIX I (continued)

Properties of Commonly Used Nonionic and Zwitterionic Detergents for Membrane Protein Solubilization and Crystallization (Compiled by M. Zulauf and H. Michel)

Generic name (short names, trade names)	Structural formula	Critical micelle concentration[a] (in H₂O)	mM	%(w/w)	Molecular weight	Suppliers[c]
Alkyl-B-D-thioglucopyranoside (thioglucoside)	C_mH_{2m+1} (sugar ring with CH_2OH, OH, OH, OH, S)	m = 7	30	0.88	294	W, C
		8	9	0.28	300	
Alkyl-β-D-maltoside	C_mH_{2m+1} (maltose, CH_2OH / OH groups)	m = 10	1.6	0.08	483	C, F
		12	0.15	0.008	511	Si
Alkanoyl-N-methyl-glucamide (MEGA-m)	$C_{m-1}H_{2m-1}{-}C\,O{-}(NCH_3){-}CH$ $-(CHOH)_4{-}CH_2OH$	m = 8	58*	1.9*	322	Se, F
		9	25*	0.9*	366	O, C, Si
		10	7*	0.25*	350	
Decanoyl-N-methyl-maltosylamin		m = 10			502	O
N-Oxides and Zwitterionic Headgroups						
N-Alkyl-N,N-dimethylamine-N-oxide (m-DAO³, Empigen OB, Ammonyx AO)	$C_mH_{2m+1}N(CH_3)_2{-}O$	m = 8	162	2.8	173	Se, O
		9	50.8	0.95	187	
		10	22.	0.42	201	
		12(LDAO)	1.4	0.03	229	
Alkyldimethylammoniopropane-sulfonate (m-DAPS, Zwittergent, SB-m)	$C_mH_{2m+1}N(CH_3)_2{-}(CH_2)_3{-}SO_3^-$	m = 8	200.	—	280	F,Se
		10	20.	1.2	308	Si, Se
		12	2.7	0.09	336	F, C
		14	0.28	0.01	364	
		16	0.03	0.001	392	

Bile Acids and Derivatives

Cholate	8*	0.3*(pH8)	431	Se, Si, C
3-(3-Cholamidopropyl)-dimethylammonio-1-propanesulfonate (CHAPS)	2—10*	0.3*	615	
3-(3-Cholamidopropyl)-dimethylammonio-2-hydroxypropane-1-sulfonate (CHAPSO)	4*	0.3*	631	
Deoxycholate	2—6*	0.3*(pH8)	415	B

Hydroxyalkyl Headgroups

Alkyl dihydroxyethyl sulfoxide	C_mH_{2m+1}–SO–$(CH_2)_2$OH	m = 8	15.8	0.35	222	B
(m-HESO, products with SO replaced by SO_2 or S are also available)					
Alkyl dihydroxypropyl sulfoxide	C_mH_{2m+1}–SO–CH_2(CHOH)CH_2OH	m = 8	20.6	0.52	238	B
(m-DOPSO, products with SO replaced by SO_2 or S are also available)					

Note: a: measured by surface tension with products used as received by M. Zulauf.
*: CMC values provided by suppliers.
c: abbreviations used.
B: Bachem Feinchemikalien AG, Hauptstrasse 144, CH-4416 Bubendorf, Switzerland.
C: Calbiochem.
F: Fluka.
N: Nikko Chemicals Co. Ltd, 1-4-8 Nihonbashi-Bakurocho, Chuoko, Tokyo 103, Japan.
O: Oxyl GmbH, Peter-Henleinstrasse 11, D-8903 Bobingen, FRG.
Se: Serva.
Si: Sigma.
W: Wako Chemicals.

APPENDIX II
LIST OF COMMERCIALLY AVAILABLE RADIOACTIVE DETERGENTS

Detergent	Supplier
Cholic acid [2.4-^{3}H]	NEN
Cholic acid [Carboxyl-^{14}C]	Am, CEA, NEN
Deoxycholic acid [Carboxyl-^{14}C]	Am, ICN
Dodecyl-β-D-maltoside [1-^{14}C]	CEA
Dodecylsodiumsulfate [^{35}S] ("SDS")	Am, CEA
$C_{12}E_8$ [1-^{14}C]	CEA
Glycocholic acid [1-^{14}C]	CEA
Lysopalmitoyl phosphatidylcholine [^{14}C]	Am, NEN
Lysostearoyl phosphatidylcholine [^{14}C]	Am
Triton X-100 [^{3}H]	NEN

Note: Octyl-β-D-glucoside and *N,N*-dimethyldodecylamine-*N*-oxide are available from CEA and American Radiolabelled Chemicals by custom synthesis.

Abbreviations: Am: Amersham International ptc; CEA: Commissariat a L'ENER-GIE ATOMIQUE, Departement de Biologie (sold in the U.S. through Research Products International); ICN: ICN Biomedicals Inc.; NEN: DuPont NEN Research Products.

APPENDIX III
SELECTED REFERENCES ON THE PREPARATION AND ANALYSIS OF TWO-DIMENSIONAL MEMBRANE PROTEIN CRYSTALS (IN HISTORICAL ORDER)

1. **Holser, W. T.**, Point groups and plane groups in a two-sided plane and their subgroups, *Kristallografiya*, 110, 266, 1958 (fundamental paper on classification and nomenclature).
2. **Seki, S., Hayashi, H., and Oda, T.**, Studies on cytochrome oxidase, *Arch. Biochem. Biophys.*, 138, 110, 1970.
3. **Vanderkooi, G., Senior, A. E., Capaldi, R. A., and Hayashi, H.**, Biological membrane structure. III. The lattice structure of membranous cytochrome oxidase, *Biochim. Biophys. Acta*, 274, 38, 1972.
4. **Unwin, P. N. T. and Henderson, R.**, Molecular structure determination by electron microscopy of unstained crystalline specimens, *J. Mol. Biol.*, 94, 425, 1975.
5. **Henderson, R. and Unwin, P. N. T.**, Three-dimensional model of purple membrane obtained by electron microscopy, *Nature*, 257, 28, 1975.
6. **Fuller, S. D., Capaldi, R. A., and Henderson, R.**, Structure of cytochrome c oxidase in deoxycholate-derived two-dimensional crystals, *J. Mol. Biol.*, 134, 305, 1979.
7. **Michel, H., Oesterhelt, D., and Henderson, R.**, Orthorhombic two-dimensional crystal form of purple membrane, *Proc. Natl. Acad. Sci. U.S.A.*, 77, 338, 1980.
8. **Kistler, J. and Stroud, R. M.**, Crystalline arrays of membrane-bound acetylcholine receptor, *Proc. Natl. Acad. Sci. U.S.A.*, 78, 3678, 1981.
9. **Amos, L. A., Henderson, R., and Unwin, P. N. T.**, Three-dimensional structure determination by electron microscopy of two-dimensional crystals, *Prog. Biophys. Molec. Biol.*, 39, 183, 1982. (Most comprehensive publication on the analysis of two-dimensional crystals.)
10. **Li, J. and Hollingshead, C.**, Formation of crystalline arrays of chlorophyll a/b-light-harvesting protein by membrane reconstitution, *Biophys. J.*, 37, 363, 1982.
11. **Corless, J. M., McCaslin, D. R., and Scott, B. L.**, Two-dimensional rhodopsin crystals from disk membranes of frog retinal rod outer segments, *Proc. Natl. Acad. Sci. U.S.A.*, 79, 1116, 1982.
12. **Hebert, H., Jørgensen, P. L., Skriver, E., and Maunsbach, A. B.**, Crystallization patterns of membrane-bound ($Na^+ + K^+$)-ATPase, *Biochim. Biophys. Acta*, 689, 571, 1982.
13. **Dux, L. and Martonosi, A.**, Two-dimensional arrays of proteins in sarcoplasmic reticulum and purified Ca^{2+}-ATPase vesicles treated with vanadate, *J. Biol. Chem.*, 258, 2599, 1983.
14. **Manella, C. A.**, Phospholipase-induced crystallization of channels in mitochondrial outer membranes, *Science*, 224, 165, 1984.
15. **Dratz, E. A., van Breemen, J. F. L., Kamps, K. M. P., Keegstra, W., and van Bruggen, E. F. J.**, Two-dimensional crystallization of bovine rhodopsin, *Biochim. Biophys. Acta*, 832, 337, 1985.
16. **Chang, C.-F., Mizushima, S., and Glaeser, R. M.**, Projected structure of the pore-forming OMP-C protein from *Escherichia coli* outer membrane, *Biophys. J.*, 47, 629, 1985.
17. **Weiss, H., Hovmöller, S., and Leonard, K.**, Preparation of membrane crystals of ubiquinol-cytochrome-c reductase from *Neurospora* mitochondria and structure analysis by electron microscopy, *Methods Enzymol.*, 126, 191, 1983.
18. **Wilkison, W. O., Walsh, J. P., Corless, J. M., and Bell, R. M.**, Crystalline arrays of the *Escherichia coli* sn-glycerol-3-phosphate acyltransferase, an integral membrane protein, *J. Biol. Chem.*, 261, 9951, 1986. (Interesting example where overproduction by genetic engineering of membrane proteins leads to *in vivo* formation of two-dimensional crystals.)
19. **Büldt, G., Mischel, M., Hentschel, M. P., Regenass, M., and Rosenbusch, J. P.**, Two-dimensional lattices of porin diffract to 6 Å resolution, *FEBS Lett.*, 205, 29, 1986.
20. **Misra, M. and Malhotra, S. K.**, Two-dimensional structure of phospholipase induced crystals of Ca^{2+}-ATPase from sarcoplasmic reticulum, *Biosci. Rep.*, 6, 1065, 1986.
21. **Lepault, J., Dargent, B., Tichelaar, W., Rosenbusch, J. P., Leonard, K., and Pattus, F.**, Three-dimensional reconstruction of maltoporin from electron microscopy and image processing, *EMBO J.*, 7, 261, 1988.
22. **Jap, B. K.**, Molecular design of PhoE porin and its functional consequences, *J. Mol. Biol.*, 205, 407, 1989.
23. **Mitra, A. K., McCarthy, M. P., and Stroud, R. M.**, Three-dimensional structure of the nicotinic acetylcholine receptor and location of the major associated 43-kD cytoskeletal protein, determined at 22 Å by low dose electron microscopy and X-ray diffraction to 12.5 Å, *J. Cell Biol.*, 109, 755, 1989.
24. **Kühlbrandt, W. and Downing, K. H.**, Two-dimensional structure of plant light-harvesting complex at 3.7 Å resolution by electron crystallography, *J. Mol. Biol.*, 207, 823, 1989.

25. **Sass, H. J., Büldt, G., Beckmann, E., Zemlin, F., van Heel, M., Zeitler, E., Rosenbusch, J. P., Dorset, D. L., and Massalski, A.**, Densely packed β-structure at the protein-lipid interface of porin is revealed by high-resolution cryo-electron microscopy, *J. Mol. Biol.*, 209, 171, 1989.
26. **Ribi, H. O., Ludwig, D. S., Merger, K. L., Schoolnik, G. K., and Kornberg, R. D.**, Three-dimensional structure of cholera toxin penetrating a lipid membrane, *Science*, 239, 1272, 1988.
27. **Ceska, T. A. and Henderson, R.**, Analysis of high-resolution biffraction patterns from purple membrane labeled with heavy-atoms, *J. Mol. Biol.*, 213, 539, 1990.
28. **Henderson, R., Baldwin, J. M., Ceska, T. A., Zemlin, F., Beckmann, E., and Downing, K. H.**, Model for the structure of Bacteriorhodopsin based on high-resolution electron cryo-microscopy, *J. Mol. Biol.*, 213, 899, 1990.

APPENDIX IV
SELECTED REFERENCES ON THE PREPARATION AND ANALYSIS OF THREE-DIMENSIONAL TYPE I MEMBRANE PROTEIN CRYSTALS (IN HISTORICAL ORDER)

1. **Vanderkooi, G., Senior, A. E., Capaldi, R. A., and Hayashi, H.,** Biological membrane structure. III. The lattice structure of membranous cytochrome oxidase, *Biochim. Biophys. Acta,* 274, 38, 1972.

2. **Henderson, R. and Shotton, D.,** Crystallization of purple membrane in three dimensions, *J. Mol. Biol.,* 139, 99, 1980.

3. **Li, J. and Hollingshead, C.,** Formation of crystalline arrays of chlorophyll a/b-light harvesting protein by membrane reconstitution, *Biophys. J.,* 37, 363, 1982.

4. **Hovmöller, S., Slaughter, M., Berriman, J., Karlsson, B., Weiss, H., and Leonard, K.,** Structural studies of cytochrome reductase, *J. Mol. Biol.,* 165, 401, 1982.

5. **Weiss, H., Hovmöller, S., and Leonard, K.,** Preparation of membrane crystals of ubiquinol-cytochrome-c reductase from *Neurospora* mitochondria and structure analysis by electron microscopy, *Methods Enzymol.,* 126, 191, 1983.

6. **Pikula, S., Mullner, N., Dux, L., and Martonosi, A.,** Stabilization and crystallization of Ca^{2+}-ATPase in detergent-solubilized sarcoplasmic reticulum, *J. Biol. Chem.,* 263, 5277, 1988.

7. **Taylor, K. A., Mullner, N., Pikula, S., Dux, L., Peracchia, C., Varga, S., and Martonosi, A.,** Electron microscope observations on Ca^{2+}-ATPase microcrystals in detergent-solubilized sarcoplasmic reticulum, *J. Biol. Chem.,* 263, 5287, 1988.

8. **Kühlbrandt, W.,** Three-dimensional crystals of the light-harvesting chlorophyll a/b protein complex from pea chloroplasts, *J. Mol. Biol.,* 194, 757, 1987.

9. **Kühlbrandt, W.,** Structure of light-harvesting chlorophyll a/b protein complex from plant photosynthetic membranes at 7 Å resolution in projection, *J. Mol. Biol.,* 202, 849, 1988.

10. **Stokes, D. L. and Green, M. N.,** Three-dimensional crystals of Ca ATPase from sarcoplasmic reticulum. Symmetry and molecular packing, *Biophys. J.,* 1, 1990.

11. **Stokes, D. L. and Green, M. N.,** Structure of Ca ATPase: electron microscopy of frozen-hydrated crystals at 6 Å resolution in projection, *J. Mol. Biol.,* 213, 529, 1990.

INDEX

Printed and bound by CPI Group (UK) Ltd, Croydon, CR0 4YY

22/10/2024

01777633-0013